Quasi Newton Algorithms Based Neural Networks in Friction Stir Welding Process

Akshansh Mishra

Founder and Project Scientific Officer

Stir Research Technologies, Uttar Pradesh, India

Dedicated to my mother Sucheta Mishra

PREFACE

When I was about to finish my undergraduate studies in Mechanical Engineering from SRM Institute of Science and Technology, I mailed few professors who excelled in Friction Stir Welding research field for a Research Assistantship. As a response I got nothing and I decided to carry out my own independent research in this field under the guidance of Professor Gopikrishna Nidigonda from SR Engineering College. Life is a struggle. In December 2017, my mom get diagnosed with an early stage Multiple Myeloma which is a type of blood cancer. Despite of this condition, my mom always motivated me to work toward my goals of becoming a good researcher and a scientist in this domain.

I founded a Research and Development firm called *Stir Research Technologies* which combines the flavours of Artificial Intelligence techniques like Neural Networks and Machine Vision into the field of Friction Stir Welding process.

This book gives an overview of the application of Neural Networks particularly trained on Quasi Newton algorithms in Friction Stir Welding research. I have tried to make this book as easy as possible so that learners will not face any type of difficulties while going through it.

Happy Learning!!

Contents

1. The Use of Artificial Neural Networks in Friction Stir Welding Research
2. Modelling Neural Networks for prefiguration of the tensile strength of Friction Stir Welded Pure Copper joints
3. Forecasting the Elongation % and Ultimate Tensile Strength of Friction Stir Welded dissimilar marine grade Aluminium alloy joints using Neural Network
4. Prediction of the mechanical properties of Friction Stir Welded Joints of aerospace alloys using Artificial Neural Network

Application of Neural Network Techniques in Friction Stir Welding process

Akshansh Mishra[1], Katyayani Jaiswal[2], A. Razal Rose[3], Gopikrishna Nidigonda[4]

[1]Founder and Project Scientific Officer, Stir Research Technologies, Uttar Pradesh-273303
[2]Department of Computer Science and Engineering, IIT Ropar, Punjab-140001
[3]Faculty of Mechanical Engineering, SRM Institute of Science and Technology, Kattangulathur
[4]Faculty of Mechanical Engineering, SR Engineering College, Warangal

Orcid id: https://orcid.org/0000-0003-4939-359X   Email id: akshansh.frictionwelding@gmail.com

Abstract: Artificial Neural Network (ANN) is a brain modelling technique by providing a new approach to computing. It introduces a less technical way to develop machine solutions. This research paper discusses the use of Artificial Neural Network (ANN) concept in Friction Stir Welding research, for example it is used in the investigation of tool parameters, for the evaluation of feedback forces which is provided by Friction Stir Welding process. Previous research also shows that ANN finds application in developing the correlation between the Friction Stir Welding parameters of the light alloy plates and mechanical properties. This method was also used for predicting average grain size in Friction Stir Welding processes.

Keywords: Friction Stir Welding; AI Technique, Artificial Neural Network; Mathematical Modelling

1. Introduction

Artificial Neural Network (ANN) can be considered as a mathematical model of a human brain. This elemental inspired method marks the next generation advancement in the computing field. The composition of Artificial Neural Network (ANN) consists of a large number of simple processing elements or basic units called *neurons*. Each neuron applies an activation function to its net input to determine its output signal. Every neuron is connected to other neurons by means of directed communication links, each with an associated weight [1]. Each neuron has an internal state called its activation level, which is a function of the inputs it has received. This can be compared with a bottle with a liquid. If we have a bottle and if we fill in the bottle with a liquid, and if we have an alarm to caution us when the level of the liquid is up to the neck of the bottle, then activation level also does the same thing as that of the alarming signal we receive. As and when the neuron receives the signal, it gets added up and when the cumulative signal reaches the activation level the neuron sends an output. Till then it keeps receiving the input. So activation level can be considered as a threshold value for us to understand.

The technique is particularly suited to problems that involve the manipulation of multiple parameters and non-linear interpolation, and as a consequence are therefore not easily amenable to conventional theoretical and mathematical approaches. Neural networks have therefore seen growing application in materials property (mechanical and physical) determination, particularly the more difficult to analyse complex multiphase and composite materials, which are growing in popularity [2].

Friction Stir is a solid state joining process developed by The Welding Institute (TWI) in the UK in 1991. This method is used for joining the alloys of aluminium, magnesium, copper, titanium and as well as steel plates [3-8]. Artificial Neural Network (ANN) plays an important role in Friction Stir Welding (FSW) Research. It is basically used to develop the applications of Friction Stir Welding (FSW) and reduce the cost of experiments. Tansel *et al* [9] represented the characteristics of Friction Stir Welding process by using Artificial Neural Network (ANN). Dehabadi *et al* [10] predicted the Vickers micro hardness of AA6061 Friction Stir Welded sheets by using Artificial Neural Network (ANN). Shojaeefard *et al* [11] performed Artificial Neural Network (ANN) analysis to model the correlation between the tool parameters (pin and shoulder diameter) and heat-affected zone, thermal, and strain value in the weld zone. Fratini *et al* [12] linked Artificial Neural Network to a finite element model (FEM) and predicted the average grain size values of butt, lap and T friction stir welded joints. Jayaraman *et al* [13] by Artificial Neural Network modelling predicted the tensile strength of A356 alloy which is a high strength Aluminium-Silicon cast alloy used in food, chemical, marine, electrical and automotive industries. This research paper mostly discusses these five papers, using them as an exemplar only to highlight the importance and use of Artificial Neural Network (ANN) in Friction Stir Welding (FSW) process.

2. How do Artificial Neural Network Works?

This is a large and complex topic because there are many different types of artificial neural network models. The most common model, which has become the foundation for most of the others, is the 3-layer fully-connected back propagation (BP) model:

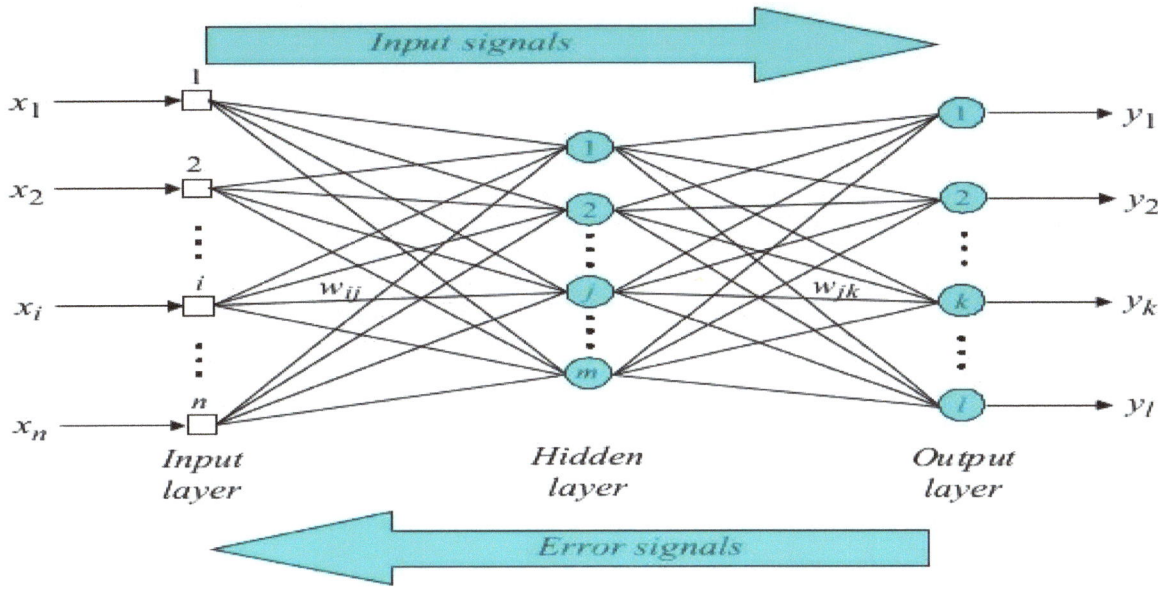

Figure 1: 3 layered fully connected back propagation model

Network Design

The basic idea is that you have three layers of "nodes." The "nodes" are intended to be analogous to neurons in a neural network of the brain, but the similarity is only metaphorical (real neurons don't work this way, but the analogy is not unreasonable). The nodes have values of 0.0 to 1.0, where 0 represents fully inactive "off" and 1 represents fully active "on" with many values in between. The three layers are an input layer, an output layer, and a "hidden" layer in the middle (hidden means neither input nor output, so not exposed to the outside world). The nodes are linked by connections which have a "weight" ("w" in the figure) that are analogous to synapses in the brain. Signal values propagate from the inputs, through the connection weights to the hidden nodes, and then onward through more connection weights to the output nodes. The number of the neurons at the first and the last layer are equal to the inputs and outputs of the ANN. The user determines the number of neurons at the intermediate layer (hidden layer) with trial and error. In most of the BP applications, each neuron is connected to all the neurons of the following layer.

In the beginning, the network will of course get the wrong answer because it knows nothing. This is where the "training" and "back propagation" comes in. The error values are propagated backward through the network using some complicated math that tells the algorithm how to modify each connection weight so that the network will get closer to the correct answer next time.

3. Use of Artificial Neural Network (ANN) in Friction Stir Welding (FSW) process

Tansel *et al* [9] used genetically optimized neural network systems (GONNS) to estimate the optimal operating condition of the friction stir welding (FSW) process. He introduced the genetically optimized neural network system (GONNS) by using Artificial Neural Network (ANN) and Genetic Algorithm (GA) together. He represented Friction Stir Welding (FSW) process in five artificial neural networks (ANN) as shown in the Figure 2. The genetically optimized neural network is shown in the Figure 3. Artificial Neural Network (ANN) is first trained by the genetically optimized neural network systems (GONNS) with experimental data. . It was observed that the inputs of the five ANNs were the same (tool rotation and welding feed rate). The estimation errors of the ANNs were better than average 0.5%. GA estimated the optimal FSW conditions to minimize or maximize one of the stir welding characteristics, while the others were kept at the desired ranges.

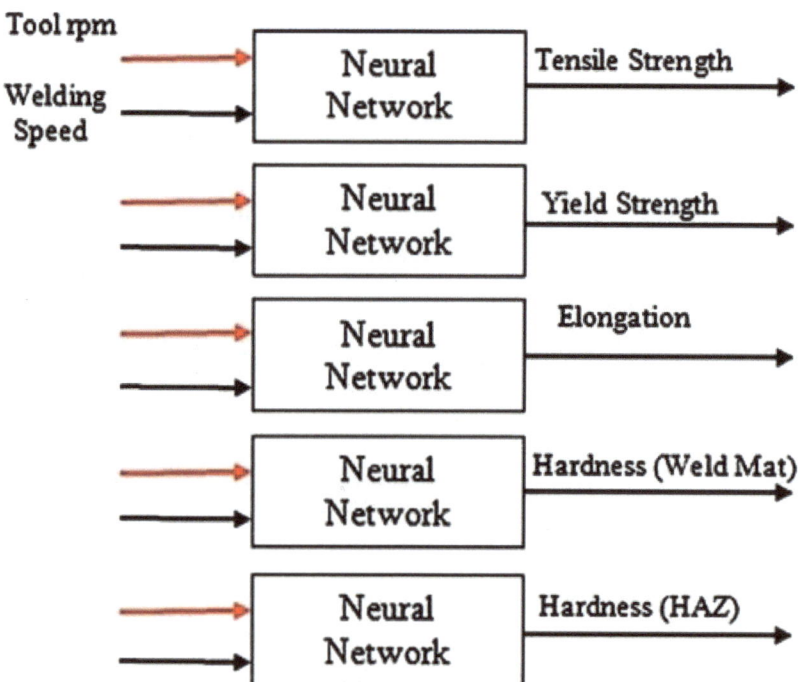

Figure 2: Five artificial neural networks for representation of the friction stir welding operation

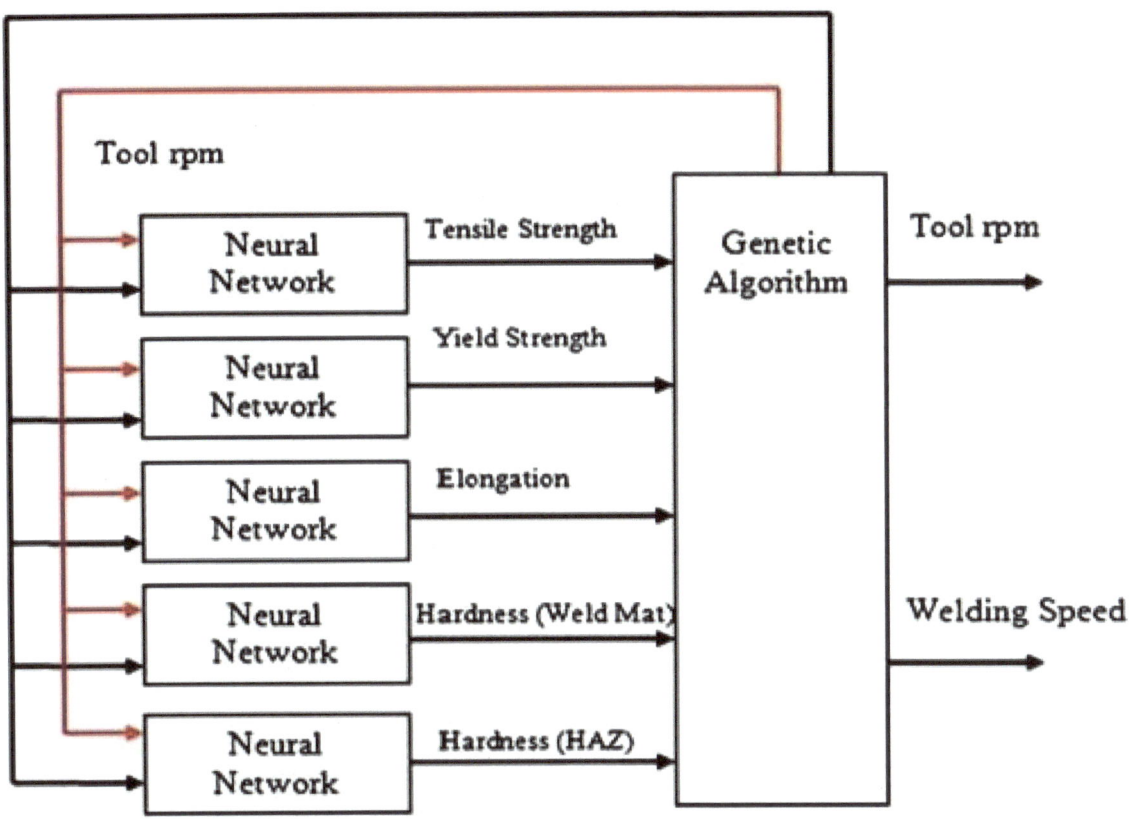

Figure 3: Genetically optimized neural network structure

Dehabadi *et al* [10] used tow Artificial Neural Network (ANN) to study the effects of thread and conical shoulder of each pin profile on the micro hardness of welded zone of AA6061 plates as shown in the Figure 4.

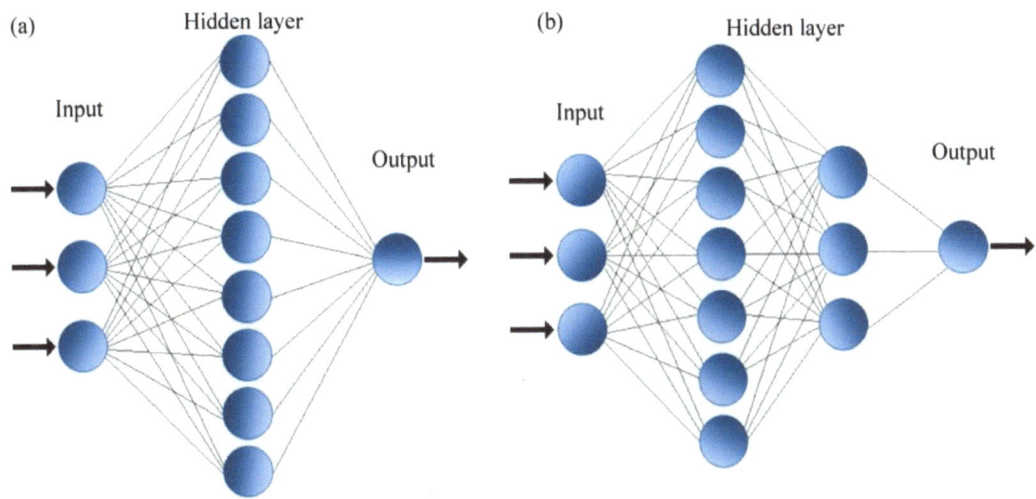

Figure 4: Schematic of neural networks in this work for predicting Vickers micro hardness in triangle (a) and tapered cylindrical (b) pin profile tools.

It was observed that the Mean absolute percentage error (MAPE) for train and test data sets did not exceed 5.4% and 7.48%, respectively. MSE values for both networks were less than 10, which indicated appropriate trained models as shown in the Figure 5.

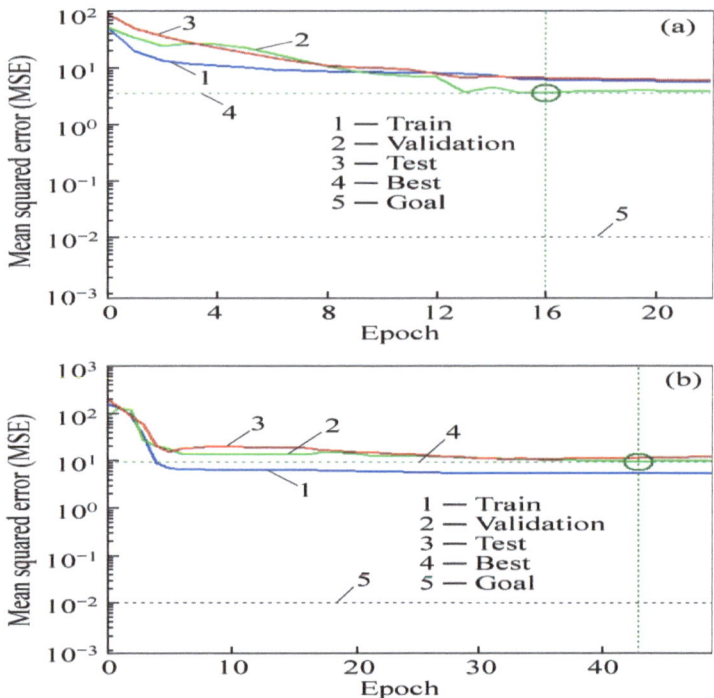

Figure 5: MSE for ANN models for triangle pin profile (a) and tapered cylindrical pin profile (b) tools

Shojaeefard *et al* [11] numerically modelled a different tool pin and shoulder diameter for a Friction Stir Welding (FSW) process. He used Feed-forward neural network with back-propagation algorithm to understand the correlation between tool dimensions and peak temperature, maximum strain, and HAZ area as shown in the Figure 6.

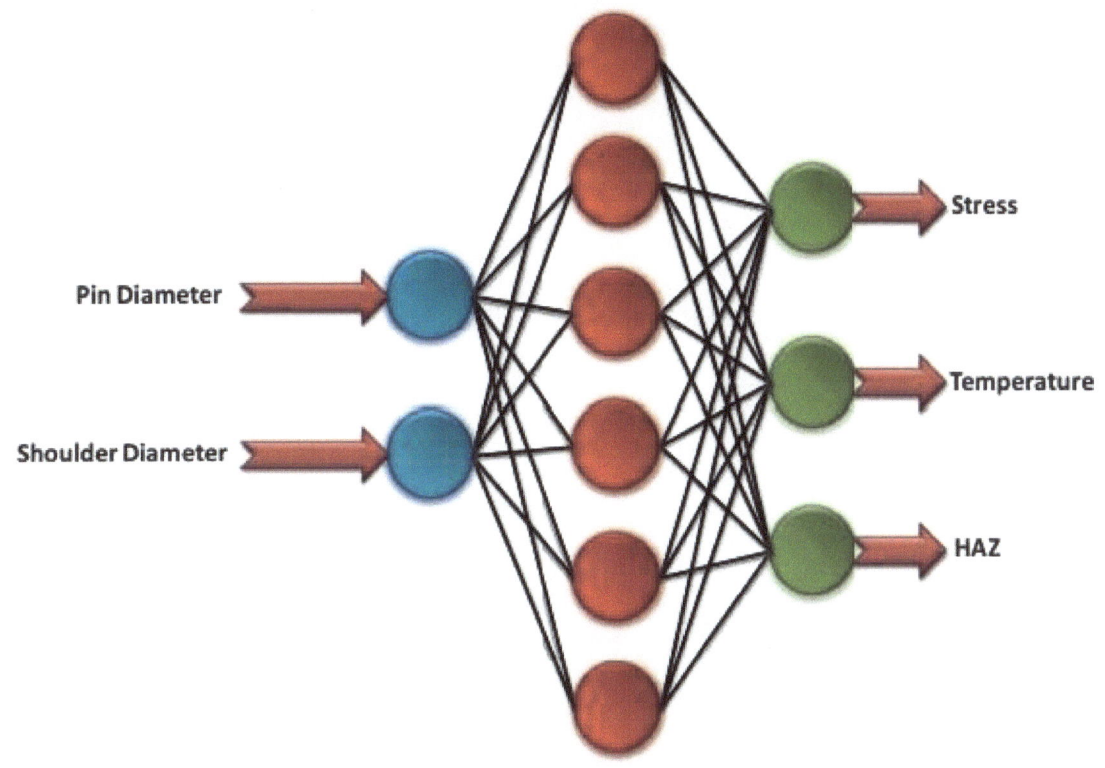

Figure 6: Three layer neural network

It can be observed that the neural network having an input layer with two neurons for each input factor (pin diameter, shoulder diameter) and an output layer with three neurons (maximum strain, maximum temperature, and HAZ area) was used. In order to get the best network architecture evaluation of several architectures were performed and trained using the experimental data. Based on this analysis, the optimal architecture was selected as 2–6–2 NN, and both activation functions in hidden layer and output layer were ''logsig.''

Fratini *et al* [12] in his research showed the capability of the AI technique in conjunction with the FE tool to predict the final microstructure of the Friction Stir Welded Joints. He designed the network architecture which was composed of five layers as shown in Figure 7. From the Figure 7 it is clearly seen that the neural network consisted of an input layer, three hidden layers and finally the output layer. Input layer was composed of four neurons which represented the local values of the equivalent plastic strain, the strain rate, temperature and the Zener Holloman parameter in a transverse section. The introduced three hidden layers have three, five and four neurons, respectively, and finally in the output layer one neuron is present corresponding to the output variable (D), namely the local value of the final average grain size. Each layer is fully connected to the next and according to the back propagation rule, the weights (w_{ij}) of the connections linking neurons belonging to two consecutive layers are adjusted in the learning stage with the aim to minimize the error between the desired output and the calculated one.

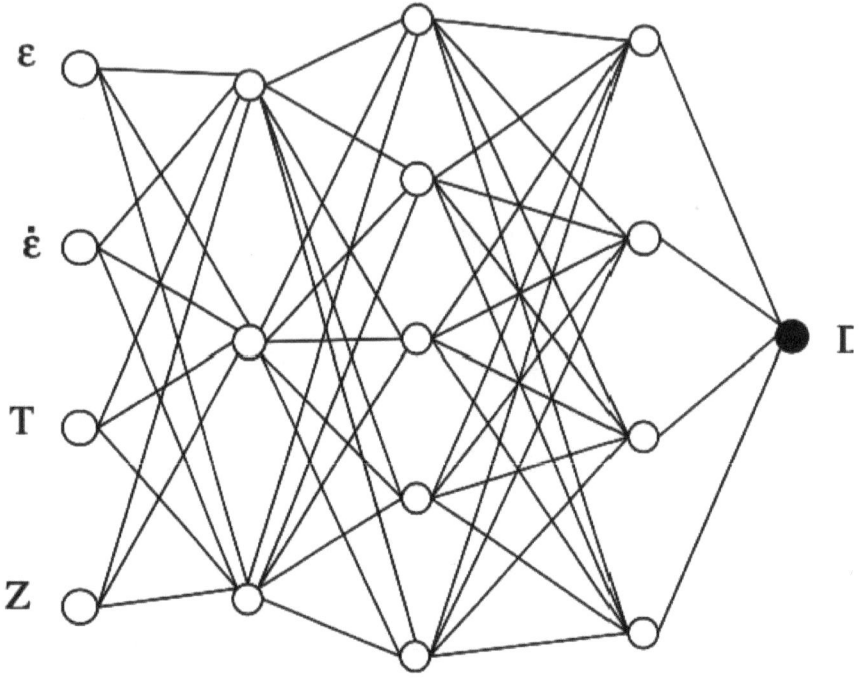

Figure 7: The utilized Neural Network architecture

Jayaraman *et al* [13] pedicted the tensile Strength of Friction Stir Welded A356 Cast Aluminium Alloy by using Response Surface Methodology (RSM) and Artificial Neural Network (ANN). He used the topology architecture of feed-forward three-layered back propagation neural network as shown in the Figure 8. He noted that the performance of Artificial Neural Network (ANNs) is better than the other techniques, especially RSM when highly non-linear behaviour is the case. Also, this technique can build an efficient model using a small number of experiments; however the technique accuracy would be better when a larger number of experiments are used to develop a model.

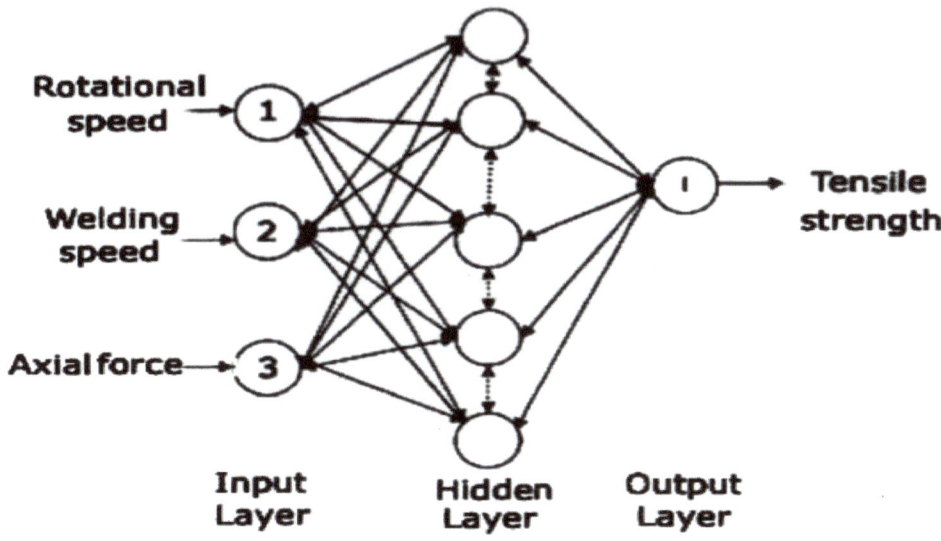

Figure 8: Used Artificial Neural Network Architecture

Maleki *et al* [14] used ANN as an efficient approach for modeling the mechanical properties of Friction Stir Welded 7075-T6 Aluminum alloy. ANN devloped was based on Back propogation algorithm. Rotational speed of tool, welding speed, axial force, shoulder diameter, pin diameter and tool hardness are regarded as inputs of the ANNs. Yield strength, tensile strength, notch-tensile strength and hardness of welding zone are gathered as outputs of neural networks as shown in the Figure 9.

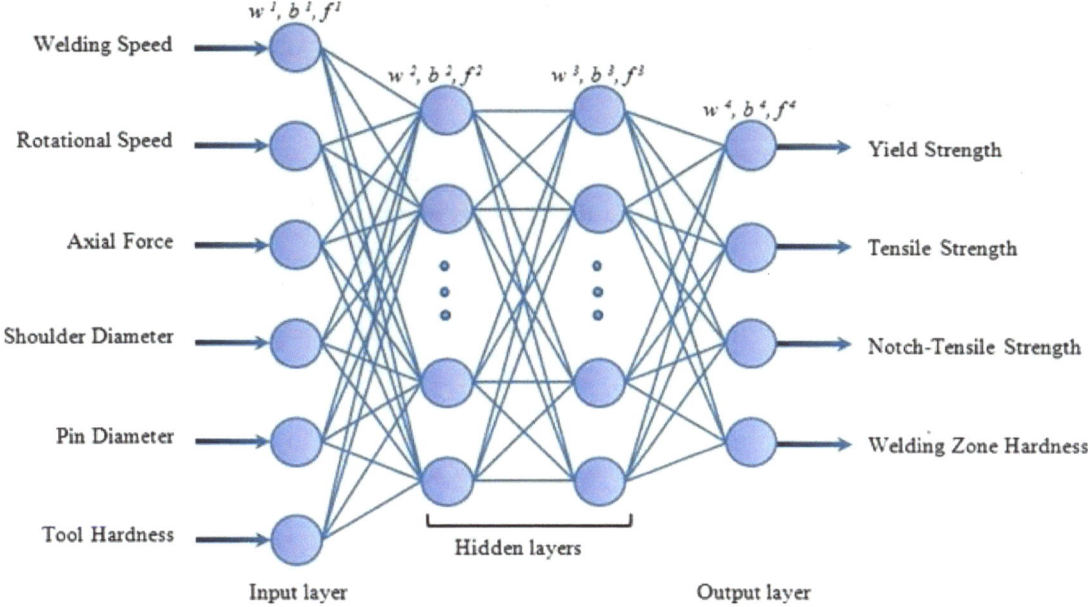

Figure 9: ANN devloped for predicting mechanical properties of FSWed 7075-T6 alloys

The least mean relative error (MRE) was obtained for the hardness of welding zone, yield strength, tensile strength and notch-tensile strength.

Khoursid *et al* [15] used the topology architecture of feed-forward three-layered back propagation neural network as illustrated in Figure 10 below for predicting the ultimate tensile strength, percentage of elongation and hardness of 6061 aluminum alloy.

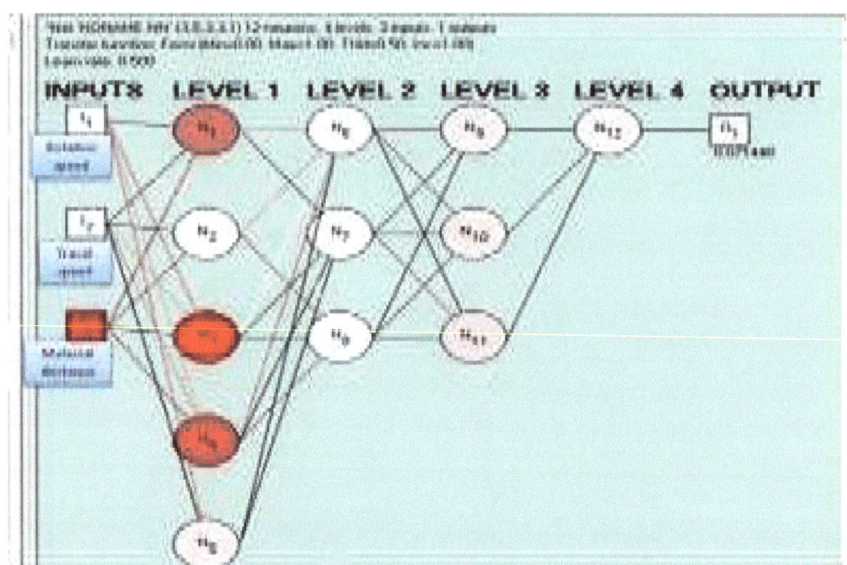

Figure 10: Propagation artificial neural network

Equation is calculated as On = F(ΣIk * Wkn). On is the neuron's output, n is the number of the neuron, Ik are the neurons inputs, k is the number of inputs, Wkn are the neurons weights. F is the Fermi function 1/(1+Exp(-4*(x-0.5))).

Software (pythia) was used for training the network model for tensile strength, the percentage of elongation and hardness prediction. The neural network described in this paper, after successful training, will be used to predict the tensile strength of friction stir welded joints of 6061 aluminum alloy within the trained range. The results obtained after training and testing On artificial neural networks are shown in the Fig.(11-13).

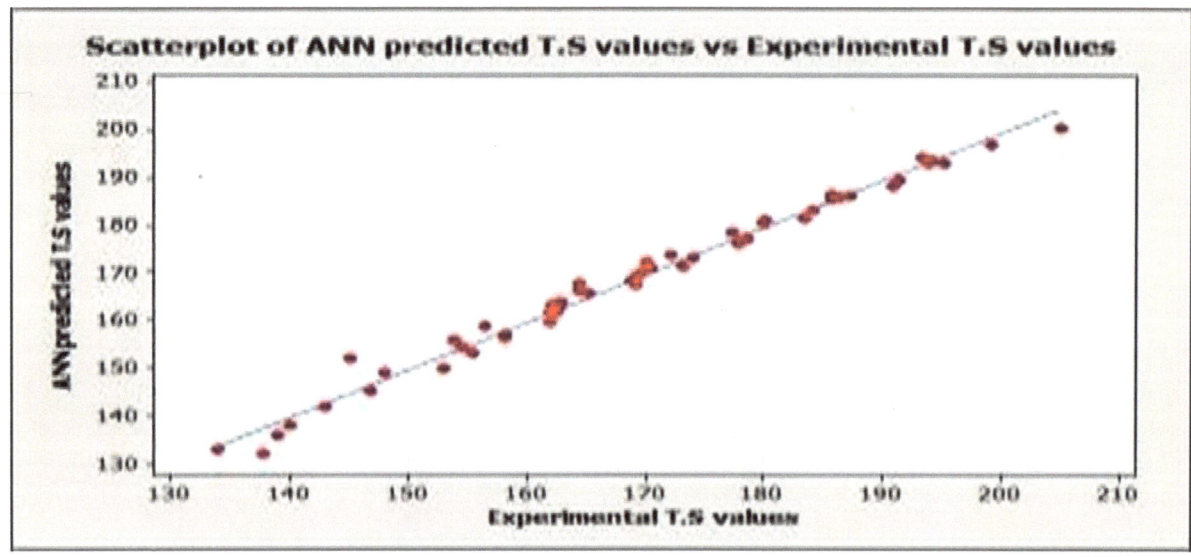

Figure 11: Relation between experimental tensile strength and predicted tensile strength

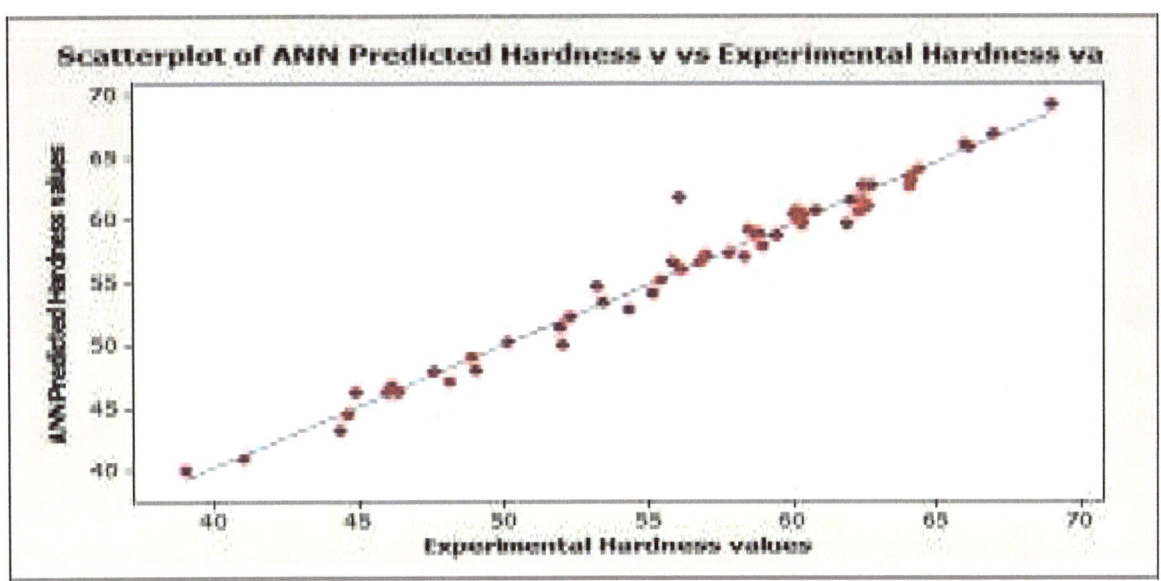

Figure 12: Relation between experimental elongation% and predicted elongation%

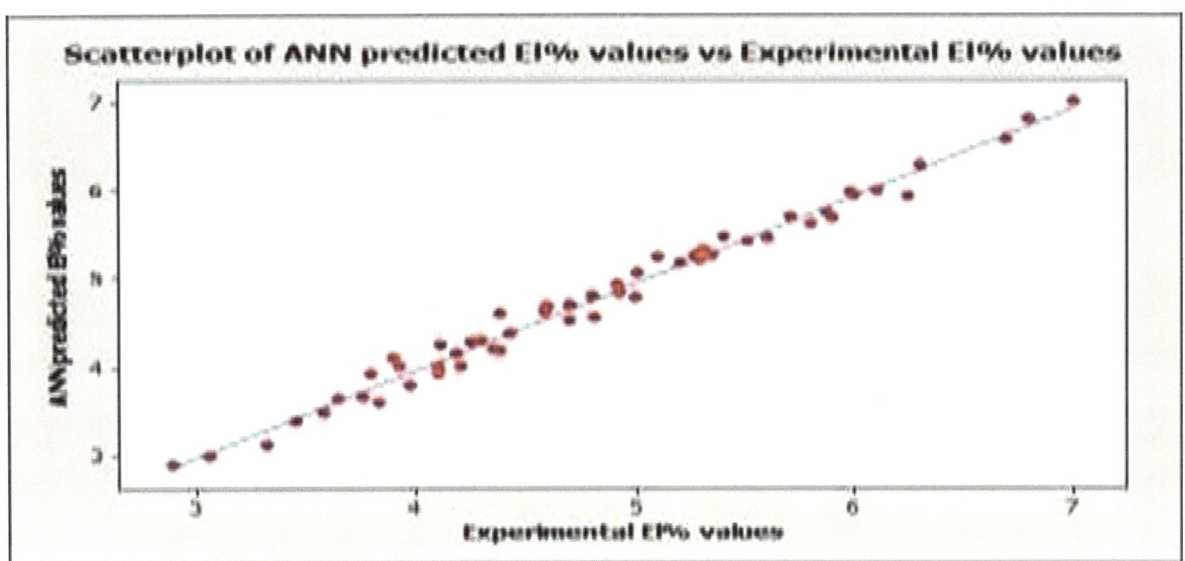

Figure 13: Relation between experimental hardness and predicted hardness

The ANN model proved to be successful in terms of agreement with experimental results ratio 96.5%.

H. Okuyucu et al. [16] developed an artificial neural network (ANN) model for the analysis and simulation of the correlation between the friction stir welding (FSW) parameters of aluminium (Al) plates and mechanical properties. The input parameters of the model consist of weld speed and tool rotation speed (TRS). The outputs of the ANN model include property parameters namely: tensile strength, yield strength, elongation, hardness of weld metal and hardness of heat effected zone (HAZ). Good performance of the ANN model was achieved. The model can be used to calculate mechanical properties of welded Al plates as functions of weld & tool rotation speeds. The combined influence of weld speed and TRS on the mechanical properties of welded Al plates was simulated. A comparison was made between measured and calculated data. The calculated results were in good agreement with measured data. The aim of the paper was to show the possibility of the use of neural networks for the calculation of the mechanical properties of welded Al plates using FSW method. Results showed that, the networks can be used as an alternative in these systems.

L Fratini and G Buffa [17] studied the continuous dynamic re-crystallisation phenomena occurring in the FSW of Al alloys. A good agreement with the experimental results was obtained using the ANN model. In regard to ANNs, it noted that ANNs perform better than the other techniques, especially RSM when highly non-linear behaviour is the case. Also, this technique can build an efficient model using a small number of experiments; however the technique accuracy would be better when a larger number of experiments are used to develop a model.

4. Conclusions

a) The Artificial Neural Network (ANN) model provides little information about the design factors and their contribution to the response if further analysis has not been done. Generation of ANN model requires a large number of iterative calculations.

b) Artificial Neural Network (ANN) method can be used to economize material and time by considering the accurate results and acceptable errors.

c) This method can be used to model the mechanical procedures. By usin g mathematical modelling methods like ANN can save time, material and costs and results are optimized in designs.

d) GONN's performance is a viable option for modelling the friction stir welding process and searching for the optimal solutions.

e) A good correlation can be observed between the predicted data obtained from the Artificial Neural Network (ANN) and FEM models.

References

1. Maind, S.B. and Wankar, P., 2014. Research paper on basic of artificial neural network. International Journal on Recent and Innovation Trends in Computing and Communication, 2(1), pp.96-100.

2. Sha, W. and Edwards, K.L., 2007. The use of artificial neural networks in materials science based research. Materials & design, 28(6), pp.1747-1752.

3. Sato, Y.S., Kokawa, H., Enomoto, M. and Jogan, S., 1999. Microstructural evolution of 6063 aluminum during friction-stir welding. Metallurgical and Materials Transactions A, 30(9), pp.2429-2437.

4. Lienert, T.J., Stellwag Jr, W.L., Grimmett, B.B. and Warke, R.W., 2003. Friction stir welding studies on mild steel. WELDING JOURNAL-NEW YORK-, 82(1), pp.1-S.

5. Mishra, Akshansh, et al. "Mechanical and Microstructure properties analysis of Friction Stir Welded Similar and Dissimilar Mg alloy joints." (2018).

6. Mishra, Rajiv S., and Z. Y. Ma. "Friction stir welding and processing." Materials science and engineering: R: reports 50.1-2 (2005): 1-78.

7. Ouyang, Jiahu, Eswar Yarrapareddy, and Radovan Kovacevic. "Microstructural evolution in the friction stir welded 6061 aluminum alloy (T6-temper condition) to copper." Journal of Materials Processing Technology 172.1 (2006): 110-122.

8. Dressler, Ulrike, Gerhard Biallas, and Ulises Alfaro Mercado. "Friction stir welding of titanium alloy TiAl6V4 to aluminium alloy AA2024-T3." Materials Science and Engineering: A 526.1-2 (2009): 113-117.

9. Tansel, Ibrahim N., et al. "Optimizations of friction stir welding of aluminum alloy by using genetically optimized neural network." The International Journal of Advanced Manufacturing Technology 48.1-4 (2010): 95-101.

10. Dehabadi, V.M., Ghorbanpour, S. & Azimi, G. J. Cent. South Univ. (2016) 23: 2146. https://doi.org/10.1007/s11771-016-3271-1

11. Shojaeefard, Mohammad Hasan, et al. "Investigation of friction stir welding tool parameters using FEM and neural network." Proceedings of the Institution of Mechanical Engineers, Part L: Journal of Materials: Design and Applications 229.3 (2015): 209-217.

12. Fratini, Livan, Gianluca Buffa, and Dina Palmeri. "Using a neural network for predicting the average grain size in friction stir welding processes." Computers & Structures 87, no. 17-18 (2009): 1166-1174.

13. Jayaraman, M., et al. "Prediction of tensile strength of friction stir welded A356 cast aluminium alloy using response surface methodology and artificial neural network." Journal for Manufacturing Science and Production 9.1-2 (2008): 45-60.

14. E Maleki 2015 IOP Conf. Ser.: Mater. Sci. Eng. 103 012034

15. A.M. Khourshid , Ahmed. M. El-Kassas , H. M. Hindawy and I. Sabry, MECHANICAL PROPERTIES OF FRICTION STIR WELDED ALUMINIUM ALLOY PIPES, European Journal of Mechanical Engineering Research, Vol.4, No.1, pp.65-78, April 2017

16. Hasan Okuyucu a, Adem Kurt a, Erol Arcaklioglu -Artificial neural network application to the friction stir welding of aluminum plates- Materials and Design 28 (2007) 78–84

17. .L Fratini and G Buffa- Continuous dynamic recrystallization phenomena modelling in friction stir welding Proceedings of the Institution of Mechanical Engineers; May 2007; 221, B5; ProQuest Science Journals pg. 857.

Modelling Neural Networks for prefiguration of the tensile strength of Friction Stir Welded Pure Copper joints

Akshansh Mishra[1], Adarsh Tiwari[2], Abhijeet Singh[3], A. Razal Rose[4]

[1]Founder and Project Scientific Officer, Stir Research Technologies, Uttar Pradesh-273303
[2]Department of Mechanical Engineering, Sagar Institute of Technology and Management, Uttar Pradesh
[3]Department of Mechanical Engineering, DIT University, Dehradun
[4]Faculty of Mechanical Engineering, SRM Institute of Science and Technology, Kattangulathur

Abstract: Artificial Neural Network (ANN) possesses a remarkable ability to extract connotation from different set of data structures. It is inspired from the mimicking of the working of biological nervous system. ANN learning abilities are more like us because they learn by examples. In this research paper prefiguration of the tensile strength of Friction Stir Welded pure copper alloys is performed. The Quasi- Newton algorithm method is used for training the neural networks. The results showed that the traverse speed is most important variable which contributes 101.3% to the output i.e. tensile strength. The accuracy of 95.71% is obtained between the actual tensile strength and predicted tensile strength.

Keywords: Artificial Neural Network; Friction Stir Welding; Tensile Strength; Quasi – Newton Algorithm

1. Introduction

ANN is a new type of structuralized model which stores information of the objects by means of the topological structures. This distributed holographic information storage instructs ANN the properties of error tolerance, error prevention, association and parallel operation. By the models of ANN which is a self- learning and non- linear mapping abilities are also obtained [1]. ANNs can be also described as the biologically inspired simulations that are performed on computer to do a certain specific set of tasks like clustering, classification, pattern recognition etc.

ANN consists of three layers i.e. Input layer, Output layer and the Hidden layer. The input layer consists of the artificial neurons which receives input from the outside world. This is the only stage where the actual learning on the network takes place. The hidden layers are incorporated between the input layers and output layers. The only job of hidden layer is to transform the input into something meaningful that the output layer unit can use in some way. The output layer consists of those artificial neurons which respond to the information that is fed into the system. It should be noted that the each of the hidden layers is individually connected to the neurons in its input layer and also to its output layer.

E. Maleki *et al* [2] used ANN technique to model the Friction Stir Welding effects on the mechanical properties of 7075-T6 Aluminum alloy. He used thirty AA7075-T6 Aluminum alloy specimens to train the neural network. The neural networks developed were based on back propagation (BP) algorithm. He considered tool rotational speed, welding speed, axial force, shoulder diameter, pin diameter and tool hardness as inputs of the ANNs while yiels strength, tensile strength, notch tensile strength and hardness of the welding zone were considered as outputs of the neural networks. He concluded that if the networks are adjusted carefully then ANN can be used for modelling of Friction Stir Welding effective parameters.

Brahma Raju *et al* [3] predicted the tensile strength of Friction Stir Welded joints of AA6061-T6 alloy by using ANN. He used three types of neural network architectures i.e. Back Propagation Neural Network (BPNN), Radial Basis Function Network (RBFN) and Generalized Regression Neural Network (GRNN). The results obtained indicated that the there was a good agreement between the experimental values and predicted values.

Dehabadi *et al* [4] investigated the ANN for predicting the Vickers microhardness of Friction Stir Welded AA6061 alloys. Mean absolute percentage error (MAPE) for train and test data sets did not exceeded 5.4% and 7.48%.

H. Okuyucu et al. [16] developed an artificial neural network (ANN) model for the analysis and simulation of the correlation between the friction stir welding (FSW) parameters of aluminium (Al) plates and mechanical properties. The input parameters of the model consist of weld speed and tool rotation speed (TRS). The outputs of the ANN model include property parameters namely: tensile strength, yield strength, elongation, hardness of weld metal and hardness of heat effected zone (HAZ). Good performance of the ANN model was achieved. The model can be used to calculate mechanical properties of welded Al plates as functions of weld & tool rotation speeds. The combined influence of weld speed and TRS on the mechanical properties of welded Al plates was simulated. A comparison was made between measured and calculated data. The calculated results were in good agreement with measured data. The aim of the paper was to show the possibility of the use of neural networks for the calculation of the mechanical properties of welded Al plates using FSW method. Results showed that, the networks can be used as an alternative in these systems.

L Fratini and G Buffa [17] studied the continuous dynamic re-crystallisation phenomena occurring in the FSW of Al alloys. A good agreement with the experimental results was obtained using the ANN model. In regard to ANNs, it noted that ANNs perform better than the other techniques, especially RSM when highly non-linear behaviour is the case. Also, this technique can build an efficient model using a small number of experiments; however the technique accuracy would be better when a larger number of experiments are used to develop a model.

In our present work ANN architectures were trained on Quasi Newton algorithm for predicting the tensile strength of Friction Stir Welded pure copper joints.

2. Experimental Procedure

The base metal used in this research was pure copper with dimensions 150 mm X 100 mm X 6mm. The plates to be joined were mounted on the fixture and Friction Stir Welding process was carried out by using H13 tool steel. The tensile specimens of the given dimensions were prepared as shown in the Figure 1. The tensile testing was carried out on Universal testing machine whose results are tabulated in the Table 1.

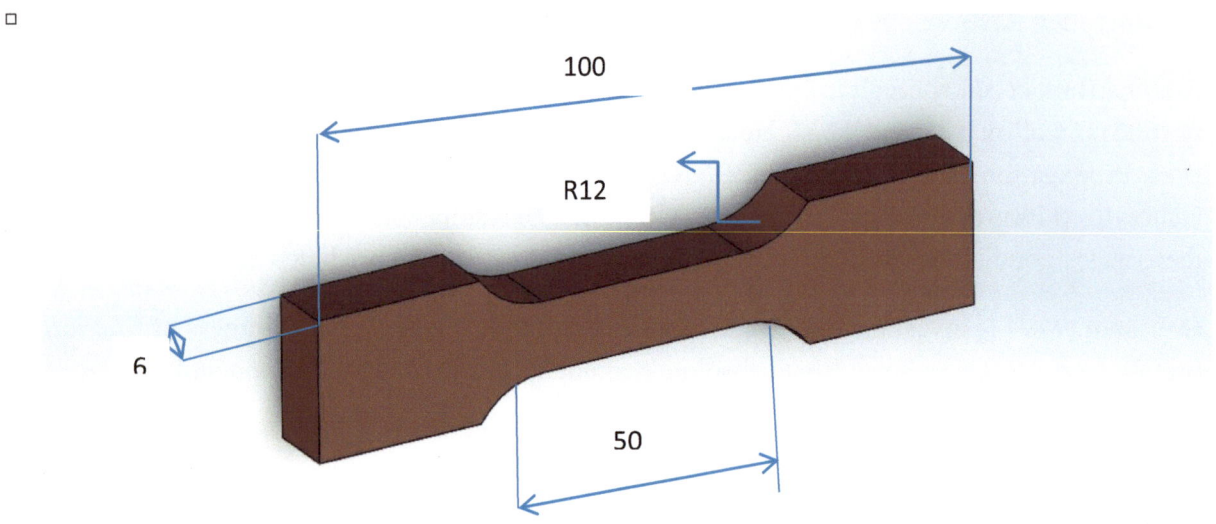

Figure 1: Design of tensile test specimen. All dimensions are in mm.

Rotational Speed in RPM	Traverse Speed (mm/min)	Tensile Strength (MPa)
1000	25	82
1000	55	62
1000	40	65
1000	10	92
2500	10	104
2500	40	77
2500	25	89
2500	55	67
4000	10	110
4000	40	82
4000	55	74
4000	25	103

Table 1: Tensile Strength at given tool rotational speed and tool traverse speed

In the present work we have used Neural Designer software for training and testing the Neural Networks. The first eleven data were trained and the tensile strength at the tool rotational speed of 4000 rpm and traverse speed of 25 mm/min was calculated by using Quasi Newton algorithm.

The various processes which were involved are shown in the form of flow chart.

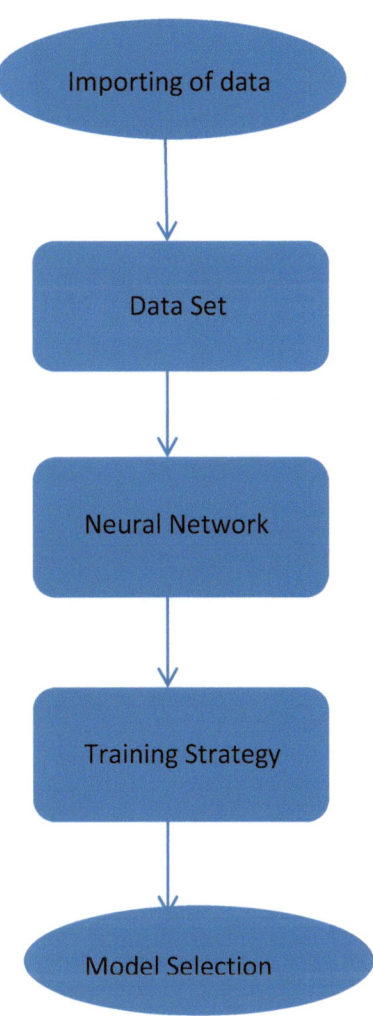

3. Results and discussions

3.1 Data Set

The data set contains the information for creating the predictive model. It comprises a data matrix in which columns represent variables and rows represent instances. Variables in a data set can be of three types: The inputs will be the independent variables; the targets will be the dependent variables; the unused variables will neither be used as inputs nor as targets. Additionally, instances can be: Training instances, which are used to construct the model; selection instances, which are used for selecting the optimal order; testing instances, which are used to validate the functioning of the model; unused instances, which are not used at all.

The next table shows a preview of the data matrix contained in the imported file Stir Research.xlsx. Here, the number of variables is 3, and the number of instances is 11.

Rotational Speed in RPM	Traverse Speed (mm/min)	Tensile Strength (MPa)
1000	25	82
1000	55	62
1000	40	65
1000	10	92
2500	10	104
2500	40	77
2500	25	89
2500	55	67
4000	10	110
4000	40	82
4000	55	74

Table 2: Imported data matrix

The following table depicts the names, units, descriptions and uses of all the variables in the data set. The numbers of inputs, targets and unused variables here are 2, 1, and 0, respectively.

Name	Use
Rotational Speed (RPM)	Input
Traverse Speed (mm/min)	Input
Tensile Strength (MPa)	Target

Table 3: Variables Table

The next chart illustrates the variables use. It depicts the numbers of inputs (2), targets (1) and unused variables (0).

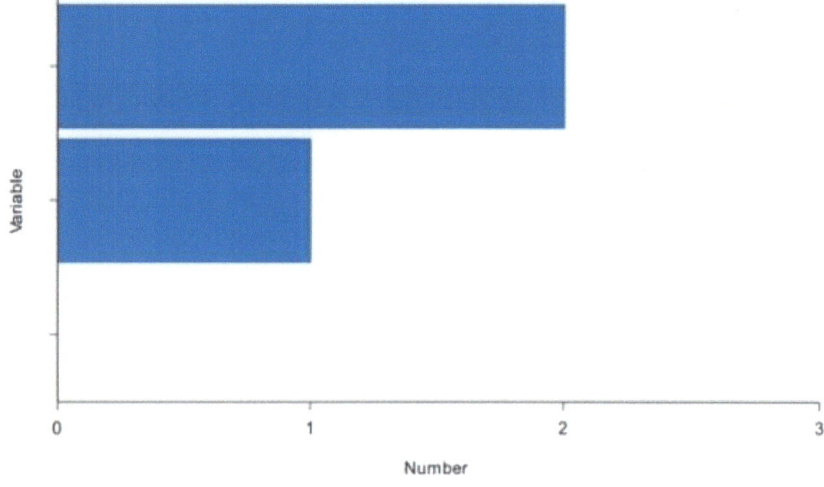

Figure 2: Variables bar charts

The following pie chart details the uses of all the instances in the data set. The total number of instances is 11. The number of training instances is 7 (63.6%), the number of selection instances is 2 (18.2%), the number of testing instances is 2 (18.2%), and the number of unused instances is 0 (0%).

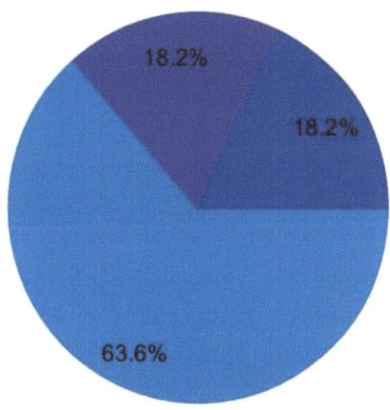

Figure 3: Instances Pie Chart

Basic statistics are very valuable information when designing a model, since they might alert to the presence of spurious data. It is a must to check for the correctness of the most important statistical measures of every single variable. The table 4 below shows the minimums, maximums, means and standard deviations of all the variables in the data set.

	Minimum	Maximum	Mean	Deviation
Rotational Speed	1000	4000	2363.64	1246.81
Traverse Speed	10	55	33.1818	18.2034
Tensile Strength	62	110	82.1818	15.5552

Table 4: Data Statistics Results

Histograms show how the data is distributed over its entire range. In approximation problems, a uniform distribution for all the variables is, in general, desirable. If the data is very irregularly distributed, then the model will probably be of bad quality. The following chart shows the histogram for the variable Rotational speed (rpm). The abscissa represents the centers of the containers, and the ordinate their corresponding frequencies. The minimum frequency is 0%, which corresponds to the bins with centers 1450, 1750, 2050, 2350, 2950, 3250 and 3550. The maximum frequency is 36.3636%, which corresponds to the bins with centers 1150 and 2650.

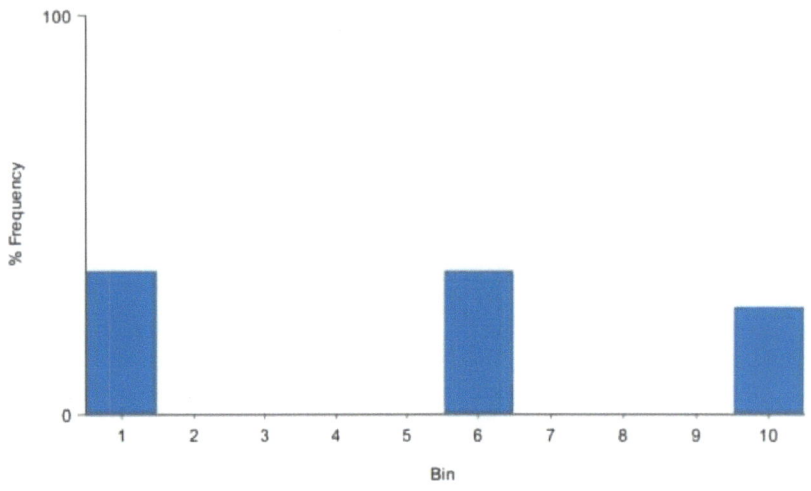

Figure 4: Rotational Speed (RPM) distribution Chart

The following chart shows the histogram for the variable Traverse Speed. The abscissa represents the centers of the containers, and the ordinate their corresponding frequencies. The minimum frequency is 0%, which corresponds to the bins with centers 16.75, 21.25, 30.25, 34.75, 43.75 and 48.25. The maximum frequency is 27.2727%, which corresponds to the bins with centers 12.25, 39.25 and 52.75.

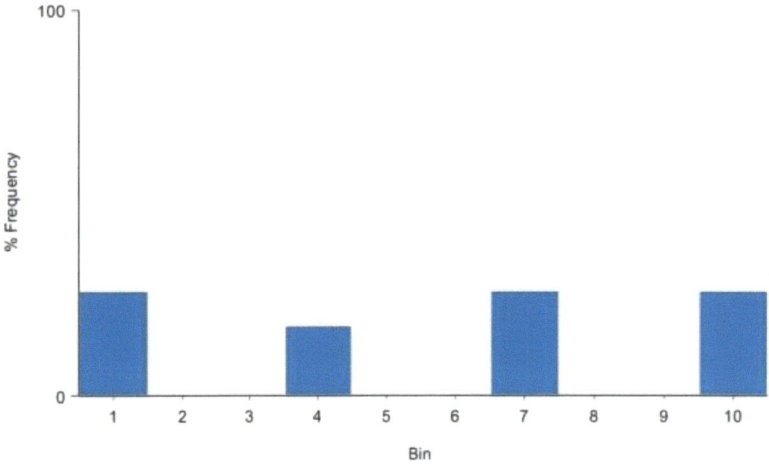

Figure 5: Traverse Speed (mm/min) distribution chart

The following chart shows the histogram for the variable Tensile Strength. The abscissa represents the centers of the containers, and the ordinate their corresponding frequencies. The minimum frequency is 0%, which corresponds to the bin with center 98. The maximum frequency is 18.1818%, which corresponds to the bins with centers 64.4 and 83.6.

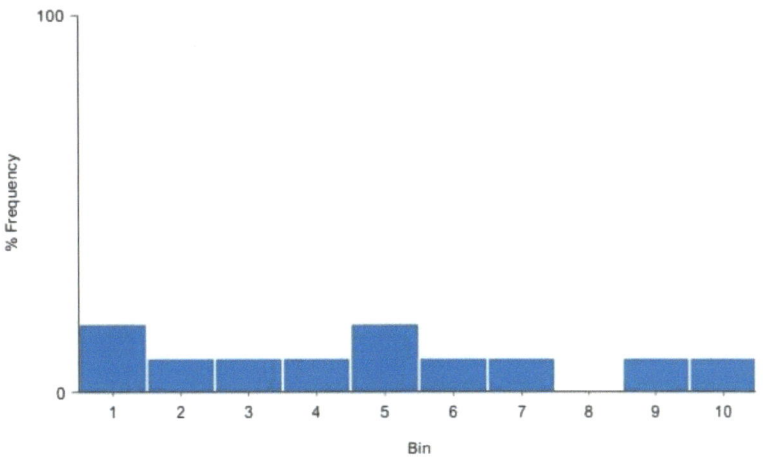

Figure 6: Tensile Strength (MPa) distribution Chart

Box plots display information about the minimum, maximum, first quartile, second quartile or median and third quartile of every variable in the data set. They consist of two parts: a box and two whiskers. The length of the box represents the interquartile range (IQR), which is the distance between the third quartile and the first quartile. The middle half of the data falls inside the interquartile range. The whisker below the box shows the minimum of the variable while the whisker above the box shows the maximum of the variable. Within the box, it will also be drawn a line which represents the median of the variable. Box plots also provide information about the shape of the data. If most of the data are concentrated between the median and the maximum, the distribution is skewed right, if most of the data are concentrated between the median and the minimum, it is said that the distribution is skewed left and if there is the same number of values at the both sides of the median, the distribution is said to be symmetric. The following chart shows the box plot for the variable Rotational speed (rpm). The minimum of the variable is 1000, the first quartile is 1000, the second quartile or median is 2500, the third quartile is 4000 and the maximum is 4000.

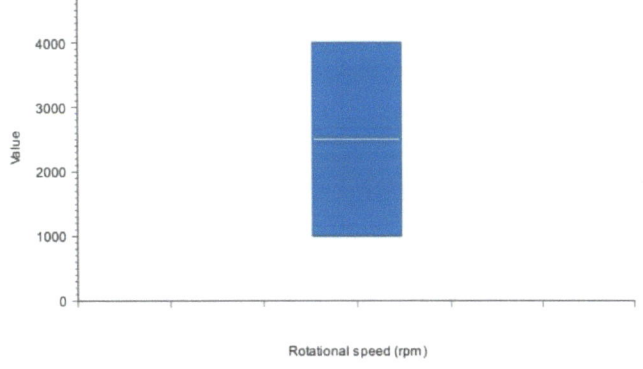

Figure 7: Rotational speed (rpm) box plot

The following chart shows the box plot for the variable Traverse Speed. The minimum of the variable is 10, the first quartile is 10, the second quartile or median is 40, the third quartile is 55 and the maximum is 55.

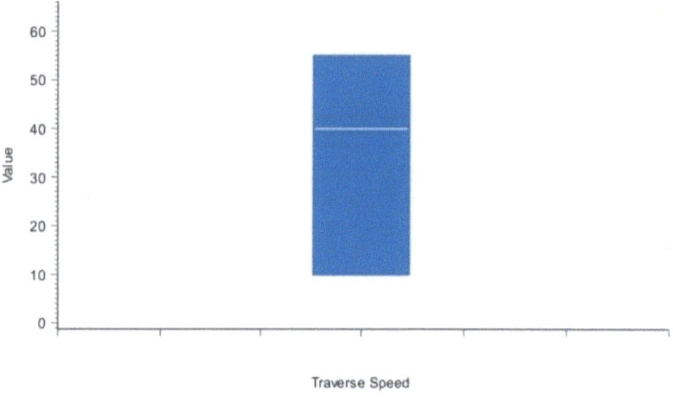

Figure 8: Traverse Speed (mm/min) box plot

The following chart shows the box plot for the variable Tensile Strength. The minimum of the variable is 62, the first quartile is 67, the second quartile or median is 82, the third quartile is 92 and the maximum is 110.

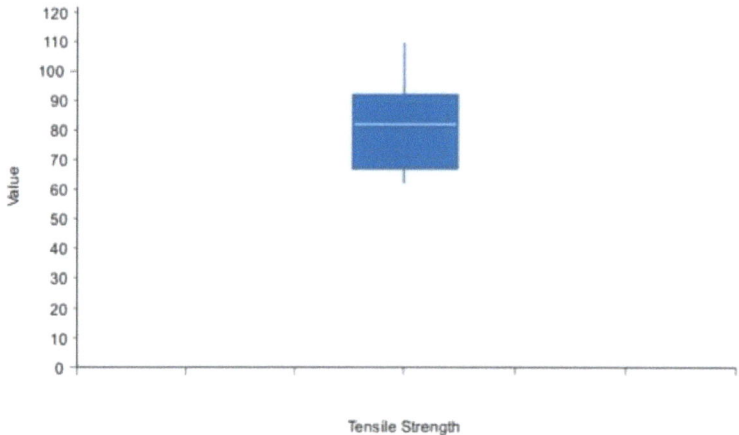

Figure 9: Tensile Strength (MPa) box plot

Target balancing task balances the distribution of targets in a data set for function regression. It unuses a given percentage of the instances whose values belong to the most populated bins. After this process, the distribution of the data will be more uniform and, in consequence, the resulting model will probably be of better quality.

The percentage of unused instances has been 10%, which corresponds to 1 instances. The following chart shows the histogram for the target variable Tensile Strength. The abscissa represents the centers of the containers, and the ordinate their corresponding frequencies. The minimum frequency is 0, which corresponds to the bins with centers 71.8, 85.3 and 98.8. The maximum frequency is 2, which corresponds to the bins with centers 67.3, 76.3 and 80.8.

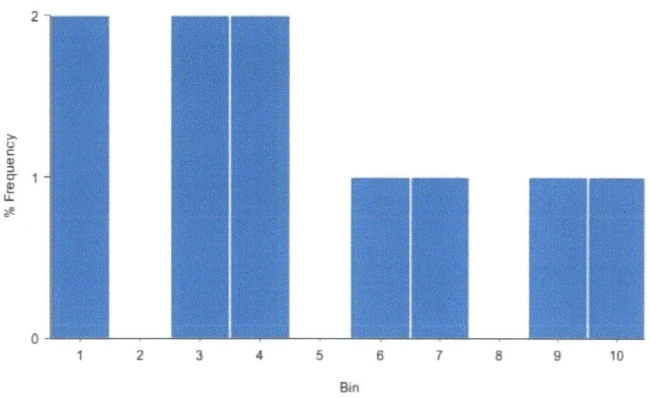

Figure 10: Tensile Strength Histogram

Scatter Plot task plots graphs of all target versus all input variables. That charts might help to see the dependencies of the targets with the inputs. The following chart shows the scatter plot for the input Rotational speed (rpm) and the target Tensile Strength.

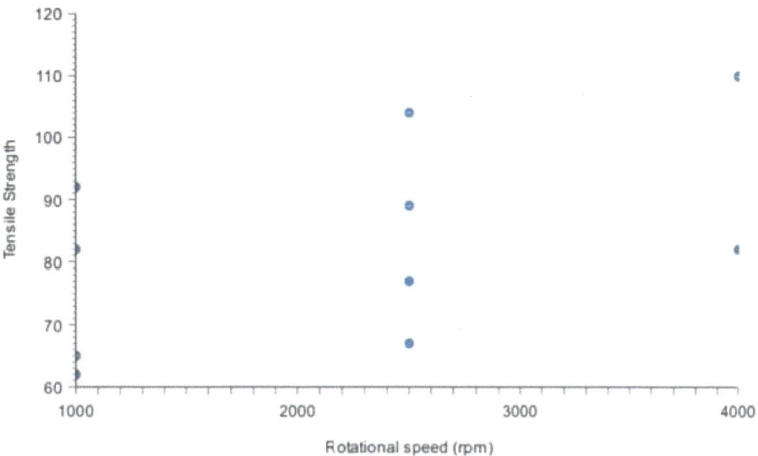

Figure 11: Tensile Strength (MPa) scatter chart vs Rotational speed (rpm)

The following chart shows the scatter plot for the input Traverse Speed and the target Tensile Strength.

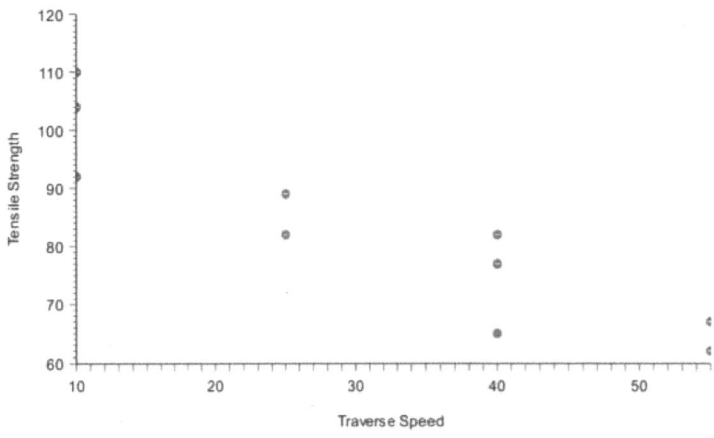

Figure 12: Tensile Strength (MPa) scatter chart vs Traverse Speed (mm/min)

Correlation Matrix task calculates the absolute values of the linear correlations among all inputs. The correlation is a numerical value between 0 and 1 that expresses the strength of the relationship between two variables. When it is close to 1 it indicates a strong relationship, and a value close to 0 indicates that there is no relationship. The following table shows the absolute value of the correlations between all input variables. The minimal correlation is 0.231869 between the variables Rotational speed and Traverse Speed. The maximal correlation is 0.231869 between the variables Rotational speed and Traverse Speed.

	Rotational Speed	**Traverse Speed**
Rotational Speed	1	0.232
Traverse Speed		1

Table 5: Correlation Matrix

It might be interesting to look for linear dependencies between single input and single target variables. This task calculates the absolute values of the correlation coefficient between all inputs and all targets. Correlations close to 1 mean that a single target is linearly correlated with a single input. Correlations close to 0 mean that there is not a linear relationship between an input and a target variables. Note that, in general, the targets depend on many inputs simultaneously. The following table shows the absolute value of the linear correlations between all input and target variables. The maximum correlation (0.895744) is yield between the input variable Traverse Speed and the target variable Tensile Strength.

	Tensile Strength
Traverse Speed	-0.896
Rotational Speed	0.365

Table 6: Tensile Strength linear correlations

The next chart illustrates the dependency of the target Tensile Strength with all the input variables.

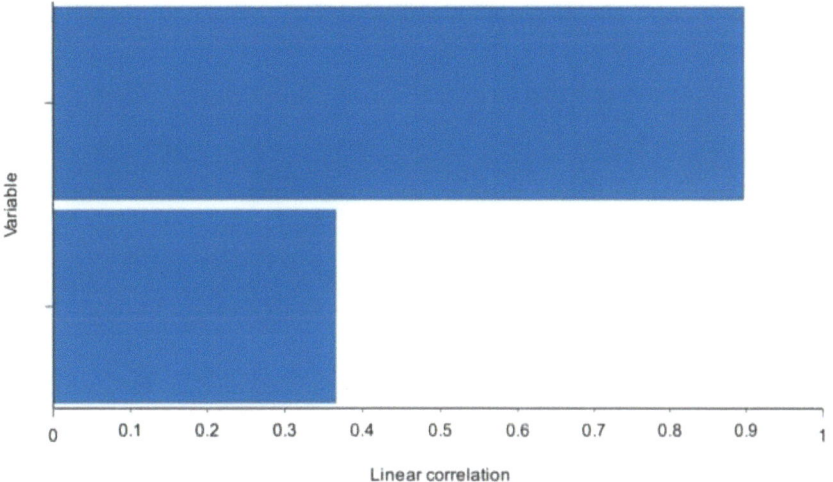

Figure 13: Tensile Strength bars chart

When designing a predictive model, the general practice is to first divide the data into three subsets. The first subset is the training set, which is used for constructing different candidate models. The second subset is the selection set, which is used to select the model exhibiting the best properties. The third subset is the testing set, which it is used for validating the final model. The following table shows the uses of all the instances in the data set. Note that the instances are arranged in rows of 10. The total number of instances is 11. The numbers of training, selection, testing and unused instances are 6, 2, 2 and 1, respectively.

	1	2	3	4	5	6	7	8	9	10
0	Train	Unused	Train	Train	Train	Sel.	Test	Test	Train	Sel.
10	Train									

Table 7: Instances Table

The following pie chart details the uses of all the instances in the data set. There are 6 instances for training (54.5%), 2 instances for selection (18.2%), 2 instances for testing (18.2%) and 1 unused instances (9.09%).

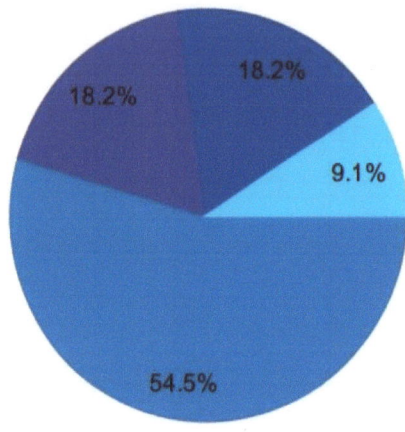

Figure 14: Instances pie chart

Principal components analysis allows to identify underlying patterns in a data set so it can be expressed in terms of other data set of lower dimension without much loss of information. The resulting data set should be able to explain most of the variance of the original data set by making a variable reduction. The final variables will be named principal components. Since this process is not reversible, it will be only applied to the input variables. The next table shows in the first column the relative explained variance for every of the principal components and in the second column the cumulative explained variance. The number of principal components of the resulting data set depends on the minimum value of the cumulative variance that it is desired the final data set had.

	Relative Variance	Cumulative Variance
1	61.7761	61.7761
2	38.2239	100

Table 8: Principal Components Results

The next chart shows the cumulative explained variance for the principal components. The x-axis represents each of the principal components and the y-axis depicts the cumulative explained variance. As it can be seen, the total explained variance for all the principal components is 100% but if the number of chosen principal components decreases also makes it the total explained variance.

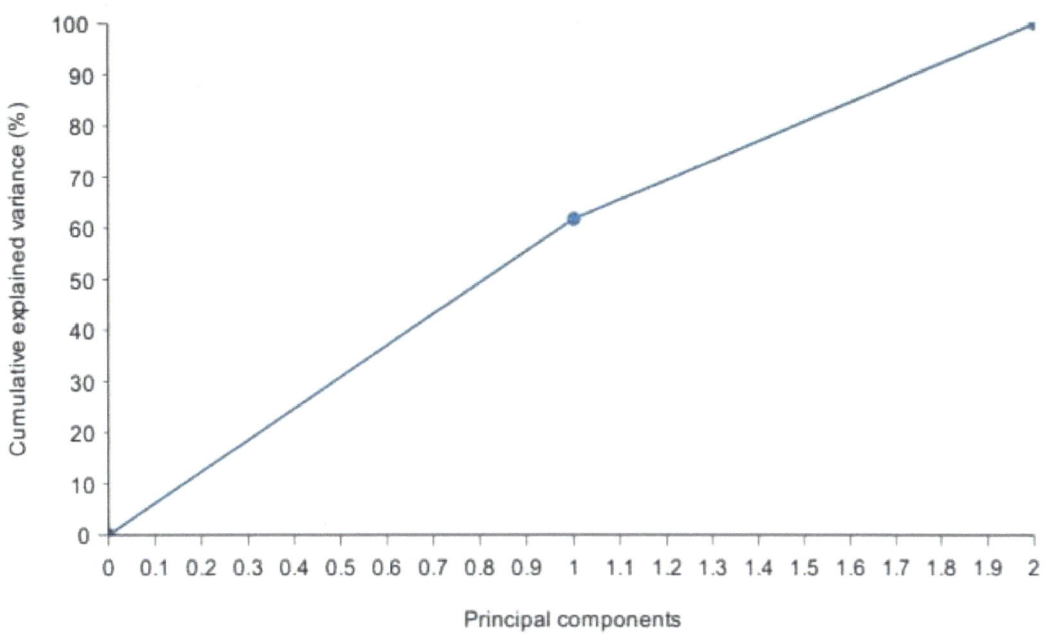

Figure 15: Explained Variance Chart

3.2 Neural Network

The neural network represents the predictive model. In Neural Designer neural networks allow deep architectures, which are a class of universal approximator. The number of inputs is 2. The next table depicts some basic information about them, including the name and the units.

	Name	Units
1	Rotational Speed	RPM
2	Traverse Speed	mm/min

Table 9: Inputs in Neural Network

The size of the scaling layer is 2, the number of inputs. The scaling method for this layer is the Minimum-Maximum. The following table shows the values which are used for scaling the inputs, which include the minimum, maximum, mean and standard deviation.

	Minimum	Maximum	Mean	Deviation
Rotational Speed	1e+003	4e+003	2.36e+003	1.25e+003
Traverse Speed	10	55	33.2	18.2

Table 10: Scaling layer values for inputs

The number of layers in the neural network is 2. The following table depicts the size of each layer and its corresponding activation function. The architecture of this neural network can be written as 2:3:1.

	Inputs Number	Neurons Number	Activation Function
1	2	3	Hyperbolic Tangent
2	3	1	Linear

Table 11: Size of each layer and their activation function

The following table shows the statistics of the parameters of the neural network. The total number of parameters is 13.

	Minimum	Maximum	Mean	Standard Devi.
Statistics	-1.19	1.41	0.103	0.786

Table 12: Statistical Parameters of the neural network

The size of the unscaling layer is 1, the number of outputs. The unscaling method for this layer is the minimum and maximum. The following table shows the values which are used for scaling the inputs, which include the minimum, maximum, mean and standard deviation.

	Minimum	Maximum	Mean	Deviation
Tensile Strength	65	110	84.2	14.8

Table 13: Values for scaling the inputs

A graphical representation of the network architecture is depicted next. It contains a scaling layer, a neural network and an unscaling layer. The yellow circles represent scaling neurons, the green circles the principal components, the blue circles perceptron neurons and the red circles unscaling neurons. The number of inputs is 2, the number of principal components is 2, and the number of outputs is 1. The complexity, represented by the numbers of hidden neurons, is 3.

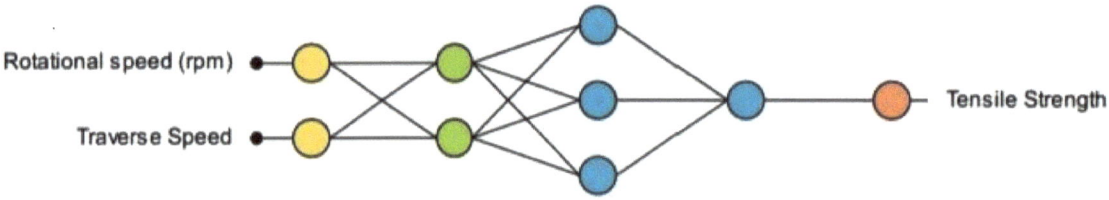

Figure 16: Neural Network architecture for predicting tensile strength

The loss index plays an important role in the use of a neural network. It defines the task the neural network is required to do, and provides a measure of the quality of the representation that it is required to learn. The choice of a suitable loss index depends on the particular application. The normalized squared error is used here as the error method. It divides the squared error between the outputs from the neural network and the targets in the data set by a normalization coefficient. If the normalized squared error has a value of unity then the neural network is predicting the data 'in the mean', while a value of zero means perfect prediction of the data. The neural parameters norm is used as the regularization method. It is applied to control the complexity of the neural network by reducing the value of the parameters. The following table shows the weight of this regularization term in the loss expression.

	Value
Natural Parameters norm weight	0.001

Table 14: Weight of the regularization term

The procedure used to carry out the learning process is called training (or learning) strategy. The training strategy is applied to the neural network in order to obtain the best possible loss. The quasi-Newton method is used here as training algorithm. It is based on Newton's method, but does not require calculation of second derivatives. Instead, the quasi-Newton method computes an approximation of the inverse Hessian at each iteration of the algorithm, by only using gradient information.

	Description	Value
Inverse Hessian approximation method	Method used to obtain a suitable training rate.	BFGS
Training rate method	Method used to calculate the step for the quasi-Newton training direction.	BrentMethod
Training rate tolerance	Maximum interval length for the training rate.	0.005
Minimum parameters increment norm	Norm of the parameters increment vector at which training stops.	1e-009
Minimum loss increase	Minimum loss improvement between two successive iterations.	1e-012
Performance goal	Goal value for the loss.	1e-012
Gradient norm goal	Goal value for the norm of the objective function gradient.	0.001
Maximum selection loss increases	Maximum number of iterations at which the selection loss increases.	100
Maximum iterations number	Maximum number of iterations to perform the training.	1000
Maximum time	Maximum training time.	3600
Reserve parameters norm history	Plot a graph with the parameters norm of each iteration.	false
Reserve loss history	Plot a graph with the loss of each iteration.	true
Reserve selection loss history	Plot a graph with the selection loss of each iteration.	true
Reserve gradient norm history	Plot a graph with the gradient norm of each iteration.	false

The following plot shows the losses in each iteration. The initial value of the training loss is 1.24264, and the final value after 96 iterations is 0.0834866. The initial value of the selection loss is 63.1754, and the final value after 96 iterations is 3224.15.

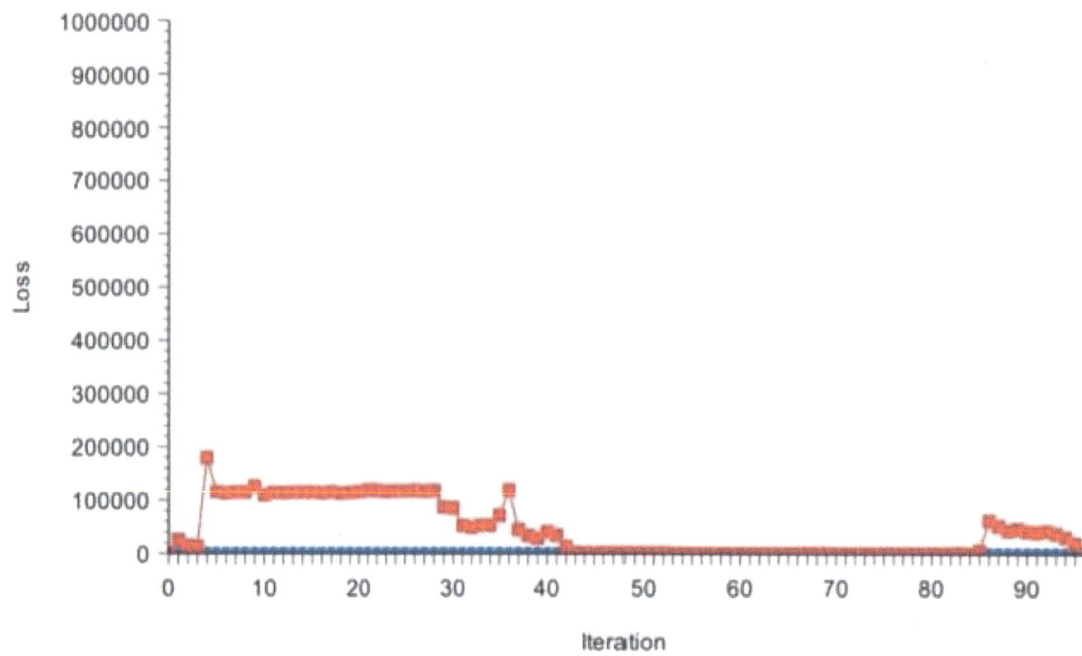

Figure 17: Loss in each iteration

The next table shows the training results by the quasi-Newton method. They include some final states from the neural network, the loss functional and the training algorithm.

	Value
Final parameters norm	80.5
Final loss	0.0835
Final selection loss	3.22e+003
Final gradient norm	0.000873
Iterations number	96
Elapsed time	00:00
Stopping criterion	Gradient norm goal

Table 15: Quasi Newton Method Results

3.3 Model Selection

Model selection is applied to find a neural network with a topology that optimizes the loss on new data. There are two different types of algorithms for model selection: Order selection algorithms and input selection algorithms. Order selection algorithms are used to find the optimal number of hidden neurons in the network. Inputs selection algorithms are responsible for finding the optimal subset of input variables.

The inputs selection algorithm chosen for this application is growing inputs. With this method, the inputs are added progressively based on their correlations with the targets.

	Description	Value
Trials number	Number of trials for each neural network.	3
Tolerance	Tolerance for the selection loss in the trainings of the algorithm.	0.01
Selection loss goal	Goal value for the selection loss.	0
Maximum selection failures	Maximum number of iterations at which the selection loss increases.	3
Maximum inputs number	Maximum number of inputs in the neural network.	2
Minimum correlation	Minimum value for the correlations to be considered.	0
Maximum correlation	Maximum value for the correlations to be considered.	1
Maximum iterations number	Maximum number of iterations to perform the algorithm.	100
Maximum time	Maximum time for the inputs selection algorithm.	3600
Plot training loss history	Plot a graph with the training losses of each iteration.	true
Plot selection loss history	Plot a graph with the selection losses of each iteration.	true

Inputs importance task calculates the selection loss when removing one input at a time. This shows which input have more influence in the outputs. The next table shows the importance of each input. If the importance takes a value greater than 1 for an input, it means that the selection error without that input is greater than with it. In the case that the importance is lower than 1, the selection error is lower without using that input. Finally, if the importance is 1, there is no difference between using the current input and not using it. The most important variable is Traverse Speed, that gets a contribution of 101.3% to the outputs.

	Contribution
Rotational Speed	0.823
Traverse Speed	1.013

Table 16: Importance of each input

The best selection is achieved by using a model whose complexity is the most appropriate to produce an adequate fit of the data. The order selection algorithm is responsible of finding the optimal number of neurons in the network. Incremental order is used here as order selection algorithm in the model selection. The next chart shows the loss history for the different subsets during the incremental order selection process. The blue line represents the training loss and the red line symbolizes the selection loss.

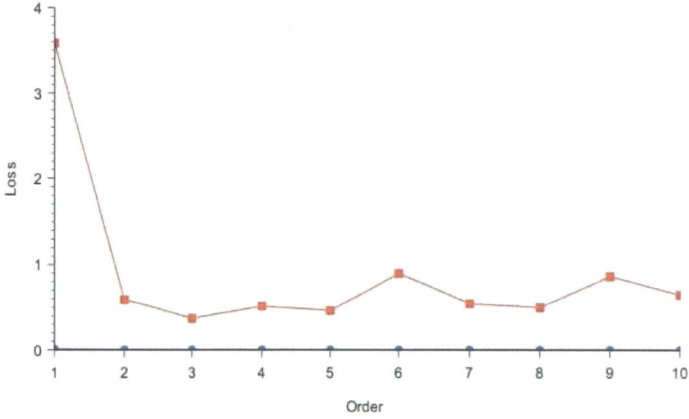

Figure 18: Incremental order loss plot

A standard method to test the loss of a model is to perform a linear regression analysis between the scaled neural network outputs and the corresponding targets for an independent testing subset. This analysis leads to 3 parameters for each output variable. The first two parameters, a and b, correspond to the y-intercept and the slope of the best linear regression relating scaled outputs and targets. The third parameter, R2, is the correlation coefficient between the scaled outputs and the targets. If we had a perfect fit (outputs exactly equal to targets), the slope would be 1, and the y-intercept would be 0. If the correlation coefficient is equal to 1, then there is a perfect correlation between the outputs from the neural network and the targets in the testing subset. The next chart illustrates the linear regression for the scaled output Tensile Strength. The predicted values are plotted versus the actual ones as squares. The coloured line indicates the best linear fit. The grey line would indicate a perfect fit.

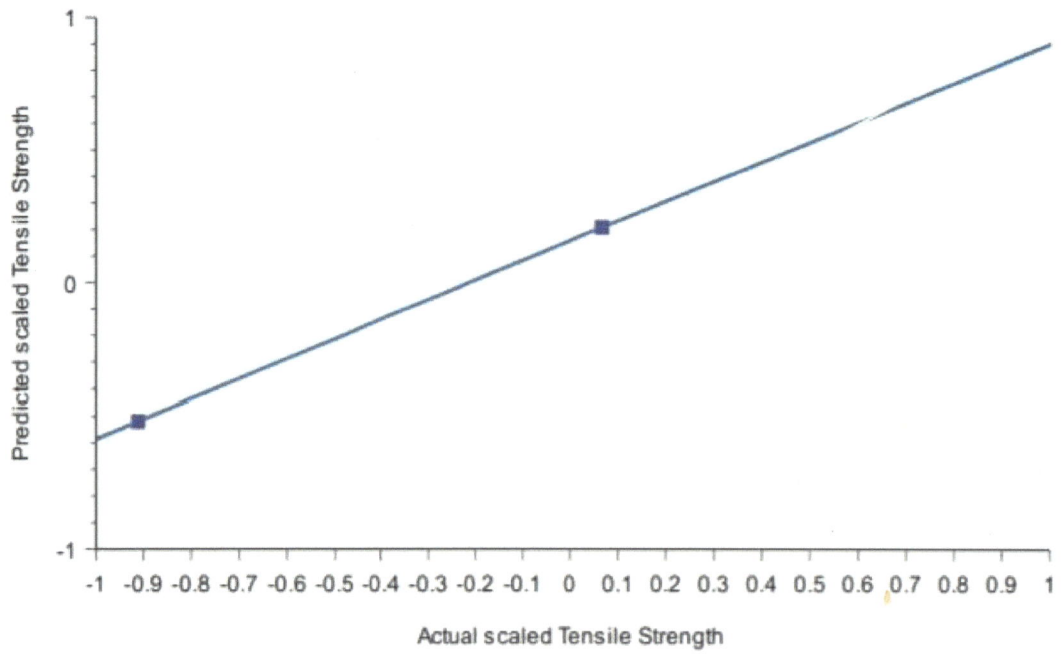

Figure 19: Tensile Strength linear regression chart

Testing error task measures all the losses of the model. it takes in account every used instance and evaluate the model for each use. The next table shows all the errors of the data for each use of them.

	Training	Selection	Testing
Sum squared error	943.666	452.014	88.1451
Mean squared error	157.278	226.007	44.0725
Root mean squared error	12.541	15.0335	6.63871
Normalized squared error	0.622129	36.1611	0.364236
Minkowski error	241.861	107.563	31.9359

Table 17: Errors table

The error data statistics measure the minimums, maximums, means and standard deviations of the errors between the neural network and the testing instances in the data set. They provide a valuable tool for testing the quality of a model. The table below shows the minimums, maximums, means and standard deviations of the absolute, relative and percentage errors of the neural network for the testing data.

	Minimum	Maximum	Mean	Deviation
Absolute error	3.18972	8.83011	6.00991	3.98835
Relative error	0.0708827	0.196225	0.133554	0.0886301
Percentage error	7.08827	19.6225	13.3554	8.86301

Table 18: Tensile strength error data statistics

A neural network produces a set of outputs for each set of inputs applied. The outputs depend, in turn, on the values of the parameters. The next table shows the input values and their corresponding output values. The input variables are Rotational speed and Traverse Speed; and the output variable is Tensile Strength.

	Value
Rotational Speed (rpm)	4000
Traverse Speed (mm/min)	25
Tensile Strength (MPa)	107.97

Table 19: Predicted Tensile Strength at given inputs

The next plot shows the output Tensile Strength as a function of the input Rotational speed (rpm). The x and y axes are defined by the range of the variables Rotational speed (rpm) and Tensile Strength, respectively.

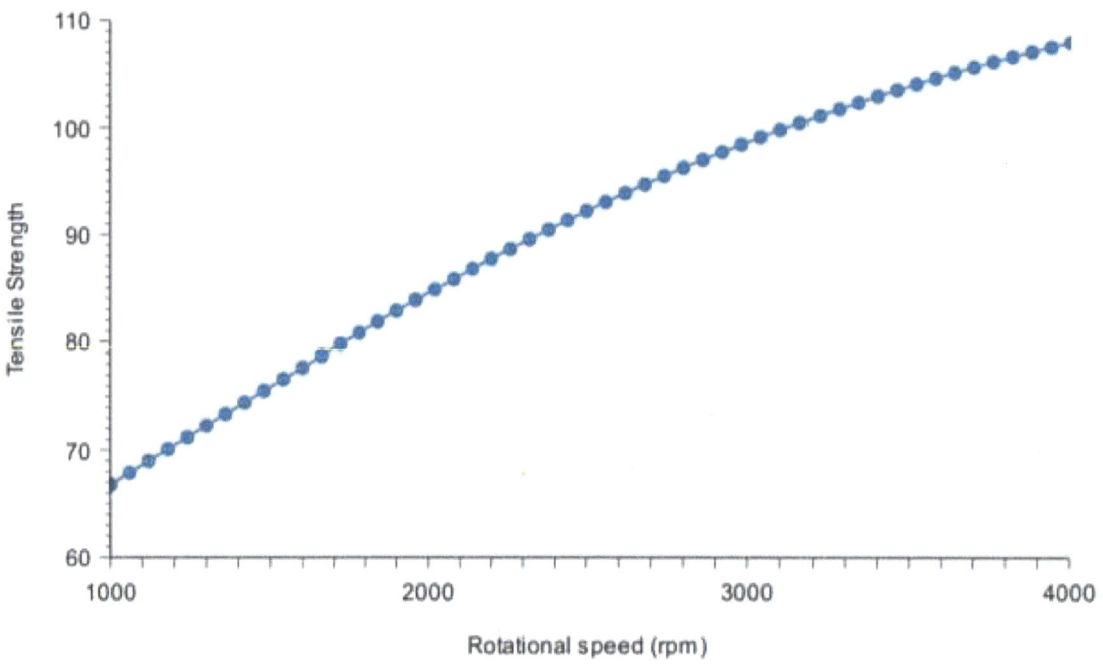

Figure 20: Tensile Strength against Rotational speed (rpm) directional line chart

The next plot shows the output Tensile Strength as a function of the input Traverse Speed. The x and y axes are defined by the range of the variables Traverse Speed and Tensile Strength, respectively.

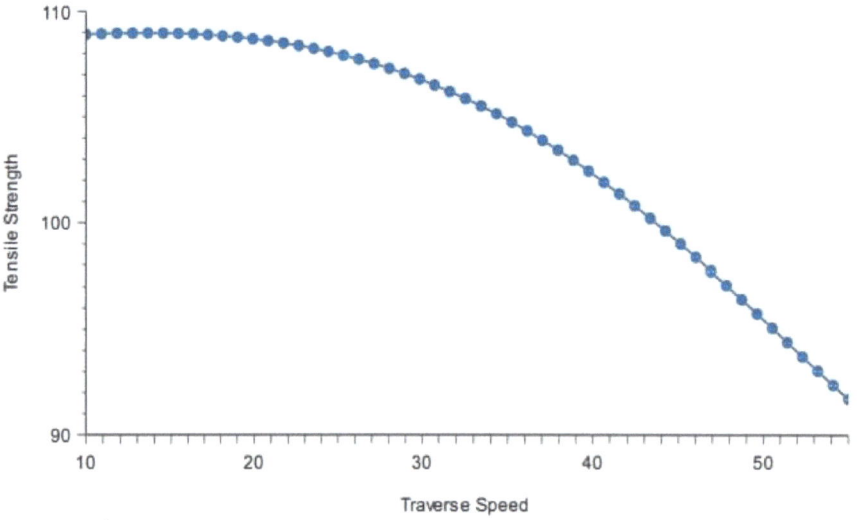

Figure 21: Tensile Strength against Traverse Speed directional line chart

4. Conclusion

The actual tensile strength calculated at the rotational speed of 4000 rpm and 25 mm/ min traverse speed is 103 MPa while predicted tensile strength from Neural Network architecture is 107.97 MPa. So the accuracy for predicting the tensile strength using the neural network architecture is 95.17%. It can be also concluded that Neural Network architecture can be used to reduce cost and time of the experiment.

References

1. Shi, Hanmin. (1997). Artificial neural network and its application in the field of mechanical engineering. 8. 5-10.

2. Maleki, E., 2015. Artificial neural networks application for modeling of friction stir welding effects on mechanical properties of 7075-T6 aluminum alloy. In IOP Conference Series: Materials Science and Engineering (Vol. 103, No. 1, p. 012034). IOP Publishing.

3. Dr. K. Brahma Raju, N. Harsha, V. K.Viswanadha Raju. Prediction Of Tensile Strength Of Friction Stir Welded Joints Using Artificial Neural Networks, International Journal of Engineering Research & Technology (IJERT), Vol. 1 Issue 9, November- 2012.

4. Dehabadi, V.M., Ghorbanpour, S. and Azimi, G., 2016. Application of artificial neural network to predict Vickers microhardness of AA6061 friction stir welded sheets. Journal of Central South University, 23(9), pp.2146-2155.

5. Okuyucu, H., Kurt, A. and Arcaklioglu, E., 2007. Artificial neural network application to the friction stir welding of aluminum plates. Materials & design, 28(1), pp.78-84.

6. Fratini, L., Buffa, G. and Palmeri, D., 2009. Using a neural network for predicting the average grain size in friction stir welding processes. Computers & Structures, 87(17-18), pp.1166-1174.

Forecasting the Elongation % and Ultimate Tensile Strength of Friction Stir Welded dissimilar marine grade Aluminium alloy joints using Neural Network

Akshansh Mishra[1]

[1]Founder and Project Scientific Officer, Stir Research Technologies, Uttar Pradesh, India

Abstract: Neural networks are a new generation of information processing paradigms designed to mimic some of the behaviours of the human brain. These networks have gained tremendous popularity due to their ability to learn, recall and generalize from training data. A number of neural network paradigms have been reported in the last four decades, and in the last decade the neural networks have been refined and widely used by researchers and application engineers. This study focuses on the prediction of the elongation % and Ultimate Tensile Strength (UTS) of the dissimilar Friction Stir Welded joints of aluminium alloys by training the Neural Network on Quasi Newton Algorithm.

Keywords: Artificial Neural Network; Quasi Newton Algorithm; Friction Stir Welding

1. Introduction

A set of processing units when assembled in a closely interconnected network, offers a surprisingly rich structure exhibiting some features of the biological neural network. Such a structure is called an artificial neural network (ANN). Since ANNs are implemented on computers, it is worth comparing the processing capabilities of a computer with those of the brain. Neural networks are slow in processing information. For the most advanced computers the cycle time corresponding to execution of one step of a program in the central processing unit is in the range of few nanoseconds. The cycle time corresponding to a neural event prompted by an external stimulus occurs in milliseconds range. Thus the computer processes information nearly a million times faster. Neural networks can perform massively parallel operations. Most programs have large number of instructions, and they operate in a sequential mode one instruction after another on a conventional computer. On the other hand, the brain operates with massively parallel operations, each of them having comparatively fewer steps. This explains the superior performance of human information processing for certain tasks, despite being several orders of magnitude slower compared to computer processing of information.

We can consider an artificial neural network (ANN) as a highly simplified model of the structure of the biological neural network. An ANN consists of interconnected processing units. The general model of a processing unit consists of a summing part followed by an output part. The summing part receives N input values, weights each value, and computes a weighted sum. The weighted sum is called the activation value. The output part produces a signal from the activation value. The sign of the weight for each input determines whether the input is excitatory (positive weight) or inhibitory (negative weight). The inputs could be discrete or continuous data values, and likewise the outputs also could be discrete or continuous. The input and output could also be deterministic or stochastic or fuzzy.

In an artificial neural network several processing units are interconnected according to some topology to accomplish a pattern recognition task. Therefore the inputs to a processing unit may come from the outputs of other processing units, and/or from external sources. The output of each unit may be given to several units including itself. The amount of the output of one unit received by another unit depends on the strength of the connection between the units, and it is reflected in the weight value associated with the connecting link. If there are N units in a given ANN, then at any instant of time each unit will have a unique activation value and a unique output value. The set of the N activation values of the network defines the activation state of the network at that instant. Likewise, the set of the N output values of the network defines the output state of the network at that instant. Depending on the discrete or continuous nature of the activation and output values, the state of the network can be described by a discrete or continuous point in an N-dimensional space.

In operation, each unit of an ANN receives inputs from other connected units and/or from an external source. A weighted sum of the inputs is computed at a given instant of time. The activation value determines the actual output from the output function unit, i.e., the output state of the unit. The output values and. other external inputs in turn determine the activation and output states of the other units. Activation dynamics determines the activation values of all the units, i.e., the activation state of the network as a function of time. The activation dynamics also determines the dynamics of the output state of the network. The set of all activation states defines the activation state space of the network. The set of all output states defines the output state space of the network. Activation dynamics determines the trajectory of the path of the states in the state space of the network. For a given network, defined by the units and their interconnections with appropriate weights, the activation states determine the short term memory function of the network.

Generally, given an external input, the activation dynamics is followed to recall a pattern stored in a network. In order to store a pattern in a network, it is necessary to adjust the weights of the connections in the network. The set of all weights on all connections in a network form a weight vector. The set of all possible weight vectors define the weight space. When the weights are changing, then the synaptic dynamics of the network determines the weight vector as a function of time. Synaptic dynamics is followed to adjust the weights in order to store the given patterns in the network. The process of adjusting the weights is referred to as learning. Once the learning process is completed, the final set of weight values corresponds to the long term memory function of the network. The procedure to incrementally update each of the weights is called a learning law or learning algorithm.

Recently, in the fields of materials joining, computer aided artificial neural network (ANN) modeling has gained increased importance. DUTTA et al [1] modeled the gas tungsten arc welding process using conventional regression analysis and neural network-based approaches and found that the performance of ANN was better compared with regression analysis. ATES et al[2] presented the use of artificial neural network for prediction of gas metal arc welding parameters. OKUYUCU et al [3] showed the possibility of the use of neural networks for the calculation of the mechanical properties of friction stir welded(FSW) aluminium plates incorporating process parameters such as rotational speed and welding speed. Tansel et al [4] used genetically optimized neural network systems (GONNS) to estimate the optimal operating condition of the friction stir welding (FSW) process. He introduced the genetically optimized neural network system (GONNS) by using Artificial Neural Network (ANN) and Genetic Algorithm (GA) together. He represented Friction Stir Welding (FSW) process in five artificial neural networks (ANN). Artificial Neural Network (ANN) is first trained by the genetically optimized neural network systems (GONNS) with experimental data. . It was observed that the inputs of the five ANNs were the same (tool rotation and welding feed rate). The estimation errors of the ANNs were better than average 0.5%. GA estimated the optimal FSW conditions to minimize or maximize one of the stir welding characteristics, while the others were kept at the desired ranges.

In our present study, we carried out Friction Stir Welding process on dissimilar alloys of Aluminium i.e. AA6061 and AA7075 alloys. The elongation % and Ultimate Tensile Strength of the dissimilar joint was predicted by training the Neural Network on Quasi Newton algorithm.

2. Experimental Procedure

In this study, 6061-T6 and 7075 aluminium alloys of the dimensions 100 mm X 50 mm X 6 mm were used as base metals. Aluminium alloy plates were machined to required dimensions for butt welding. For Friction Stir Welding H13 tool steel with chemical composition 0.406% C, 1.096% Si, 0.443% Mn, 4.952% Cr, 1.251% Mo, 0.183% V with given dimensions was used as weld tool. The assessed mechanical properties are tabulated in the Table 1.

Table 1: Assessed Mechanical properties

Exp. No	TRS (rpm)	WS (mm/min.)	YS (MPa)	UTS (MPa)	E %
1	400	10	137.20	145.51	3.52
2	400	20	135.10	143.28	3.21
3	400	30	132.44	140.46	3.00
4	400	40	130.27	138.16	2.73
5	600	10	146.70	161.14	4.22
6	600	20	144.50	158.70	3.85
7	600	30	141.61	155.55	3.60
8	600	40	139.30	153.01	3.27
9	800	10	156.80	184.10	5.92
10	800	20	154.50	181.40	4.49
11	800	30	151.36	177.70	4.20
12	800	40	148.88	174.80	3.82
13	1000	10	167.30	203.30	5.63
14	1000	20	164.80	204.25	5.13
15	1000	30	161.50	198.20	5.80
16	1000	40	158.85	145.50	4.36

Microsoft Excel file was created for first 15 datasets in order to train the Neural Network. Neural Designer software is used for importing the Excel file and further training and testing datasets. Tool rotational speed (rpm) and Welding speed (WS) are the inputs and Elongation % and Ultimate Tensile Strength (UTS) are targets in the Neural Network architecture as shown in the Figure 1.

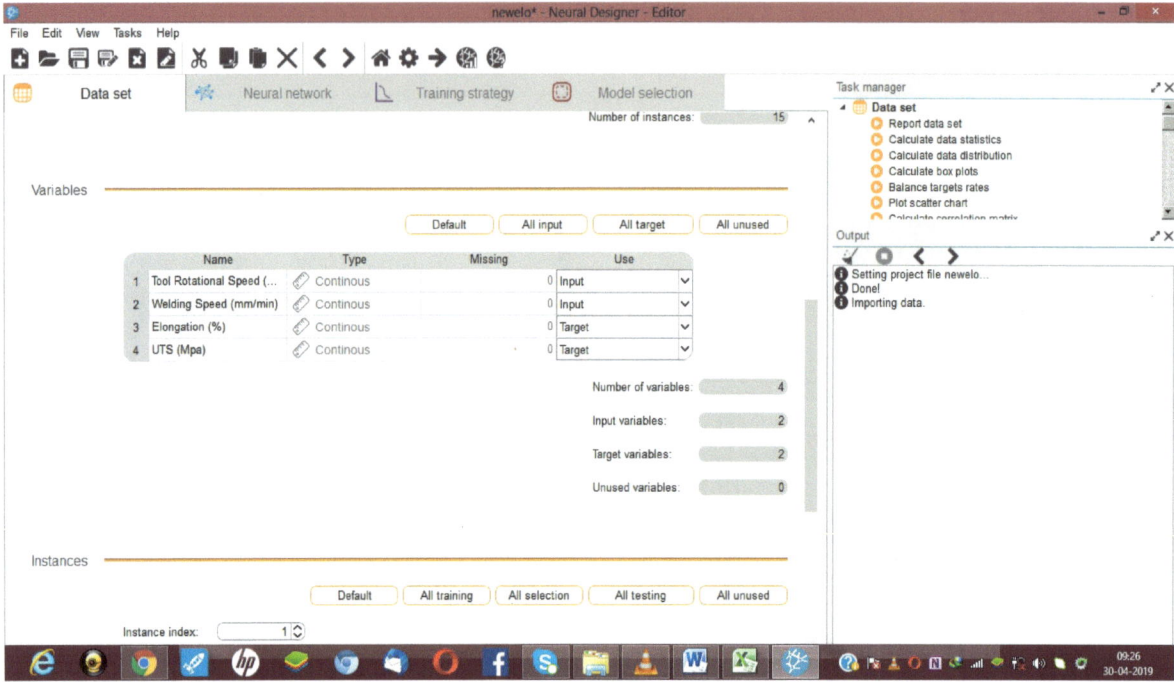

Figure 1: Neural Designer software workbench

3. Results and Discussions

3.1 Data Sets

The data set contains the information for creating the predictive model. It comprises a data matrix in which columns represent variables and rows represent instances. Variables in a data set can be of three types: The inputs will be the independent variables; the targets will be the dependent variables; the unused variables will neither be used as inputs nor as targets. Additionally, instances can be: Training instances, which are used to construct the model; selection instances, which are used for selecting the optimal order; testing instances, which are used to validate the functioning of the model; unused instances, which are not used at all. The next table shows a preview of the data matrix contained in the file elongation % prediction.xlsx. Here, the number of variables is 4, and the number of instances is 15.

Table 2: Data preview table

	Tool Rotational Speed (RPM)	Welding Speed (mm/min)	Elongation (%)	UTS (Mpa)
1	400	10	3.52	145.51
2	400	20	3.21	143.28
...				
15	1000	30	5.8	198.2

The following table depicts the names, units, descriptions and uses of all the variables in the data set. The numbers of inputs, targets and unused variables here are 2, 2, and 0, respectively.

40

Table 3: Variables Table

	Name	Use
1	Tool Rotational Speed (RPM)	Input
2	Welding Speed (mm/min)	Input
3	Elongation (%)	Target
4	UTS (Mpa)	Target

The next chart illustrates the variables use. It depicts the numbers of inputs (2), targets (2) and unused variables (0).

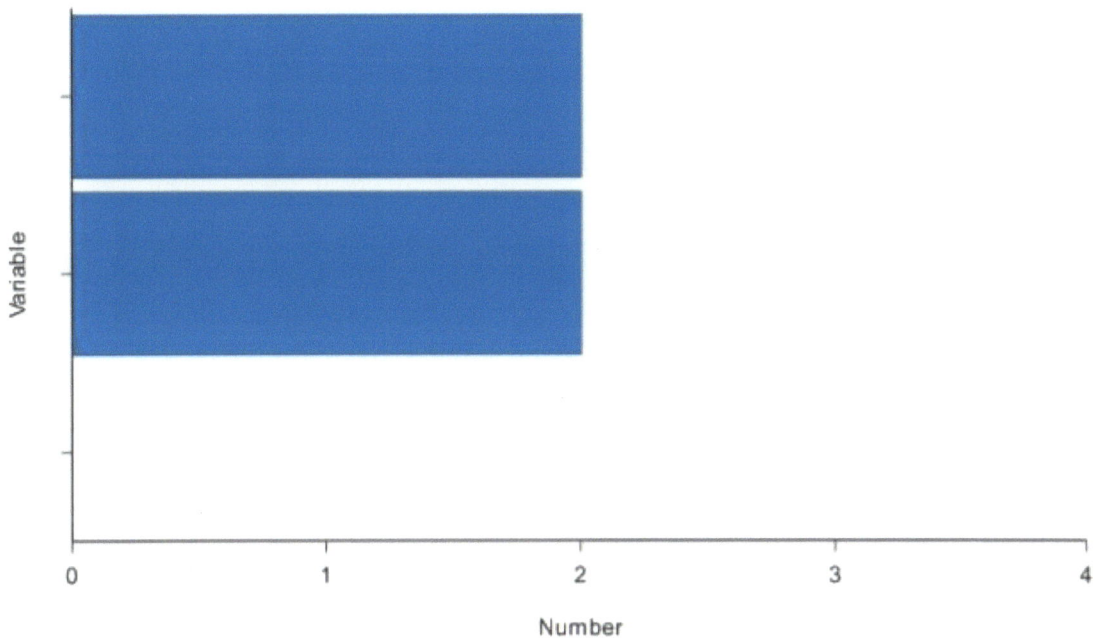

Figure 3: Variables bars chart

The following pie chart details the uses of all the instances in the data set. The total number of instances is 15. The number of training instances is 9 (60%), the number of selection instances is 3 (20%), the number of testing instances is 3 (20%), and the number of unused instances is 0 (0%).

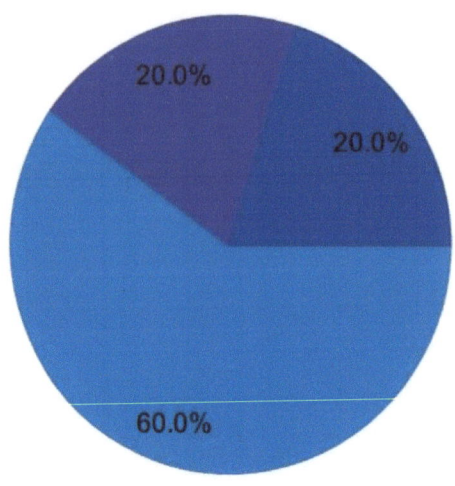

Figure 4: Instances pie chart

There are not missing values in the data set. Basic statistics are very valuable information when designing a model, since they might alert to the presence of spurious data. It is a must to check for the correctness of the most important statistical measures of every single variable. The table below shows the minimums, maximums, means and standard deviations of all the variables in the data set.

Table 4: Data statistics results

	Minimum	Maximum	Mean	Deviation
Tool Rotational Speed (RPM)	400	1000	680	224.245
Welding Speed (mm/min)	10	40	24	11.2122
Elongation (%)	2.73	5.92	4.15933	1.03589
UTS (Mpa)	138.16	204.25	167.971	22.8821

Histograms show how the data is distributed over its entire range. In approximation problems, a uniform distribution for all the variables is, in general, desirable. If the data is very irregularly distributed, then the model will probably be of bad quality. The following chart shows the histogram for the variable Tool Rotational Speed (RPM). The abscissa represents the centers of the containers, and the ordinate their corresponding frequencies. The minimum frequency is 0%, which corresponds to the bins with centers 490, 550, 670, 730, 850 and 910. The maximum frequency is 26.6667%, which corresponds to the bins with centers 430, 610 and 790.

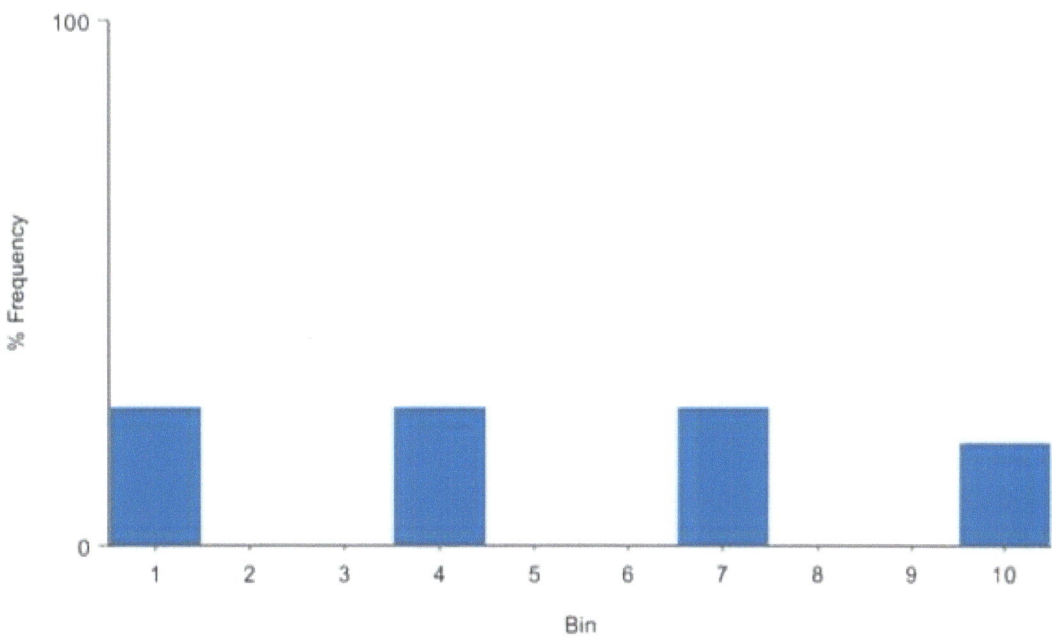

Figure 5: Tool Rotational Speed (RPM) distribution

The following chart shows the histogram for the variable Welding Speed (mm/min). The abscissa represents the centers of the containers, and the ordinate their corresponding frequencies. The minimum frequency is 0%, which corresponds to the bins with centers 14.5, 17.5, 23.5, 26.5, 32.5 and 35.5. The maximum frequency is 26.6667%, which corresponds to the bins with centers 11.5, 20.5 and 29.5.

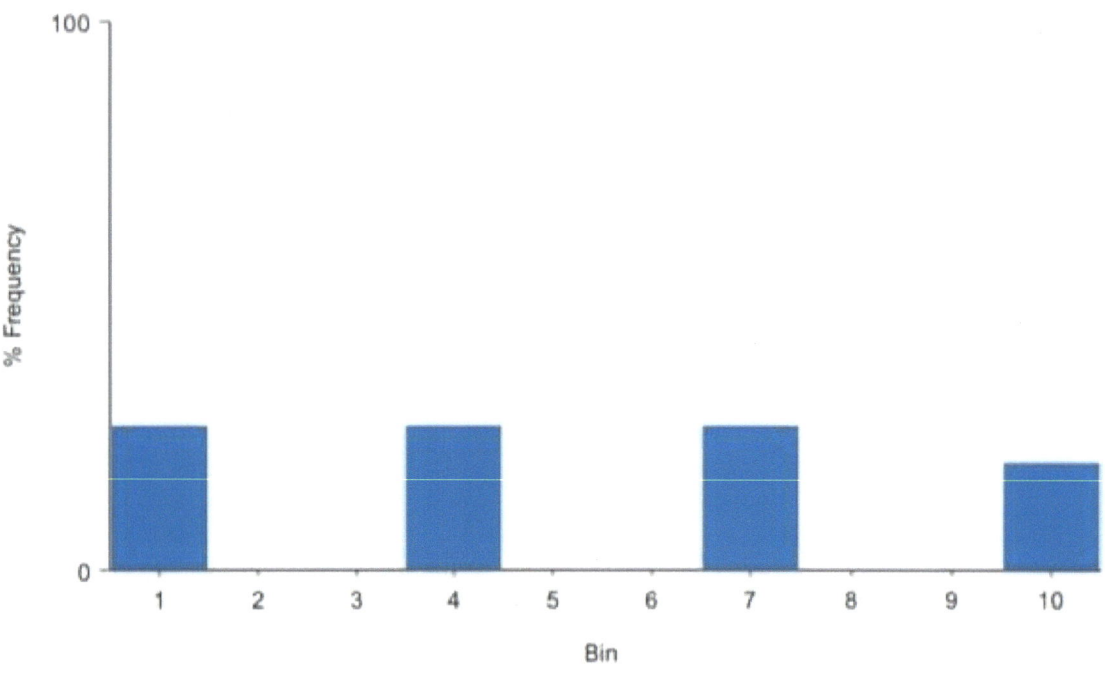

Figure 6: Welding Speed (mm/min) distribution

The following chart shows the histogram for the variable Elongation (%). The abscissa represents the centers of the containers, and the ordinate their corresponding frequencies. The minimum frequency is 0%, which corresponds to the bins with centers 4.803 and 5.441. The maximum frequency is 20%, which corresponds to the bin with center 5.761.

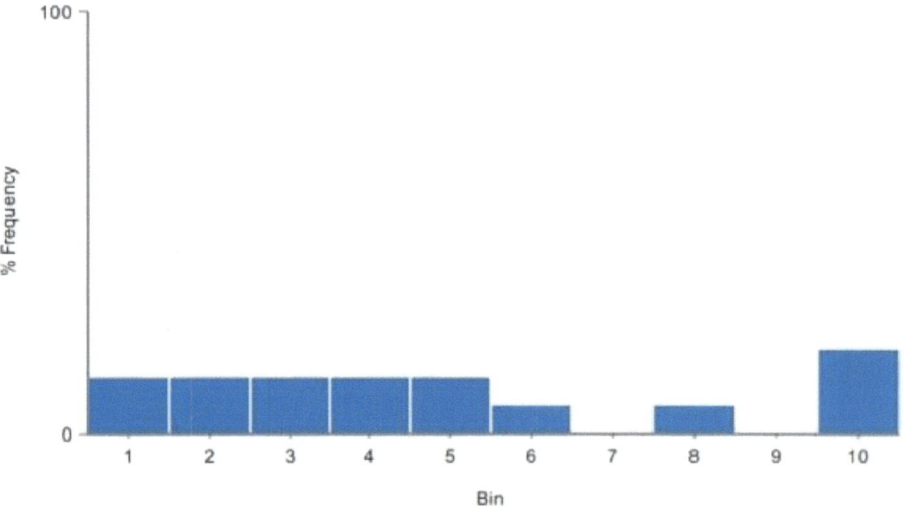

Figure 7: Elongation (%) distribution

The following chart shows the histogram for the variable UTS (Mpa). The abscissa represents the centers of the containers, and the ordinate their corresponding frequencies. The minimum frequency is 0%, which corresponds to the bins with centers 167.9, 187.7 and 194.3. The maximum frequency is 20%, which corresponds to the bins with centers 141.5 and 200.9.

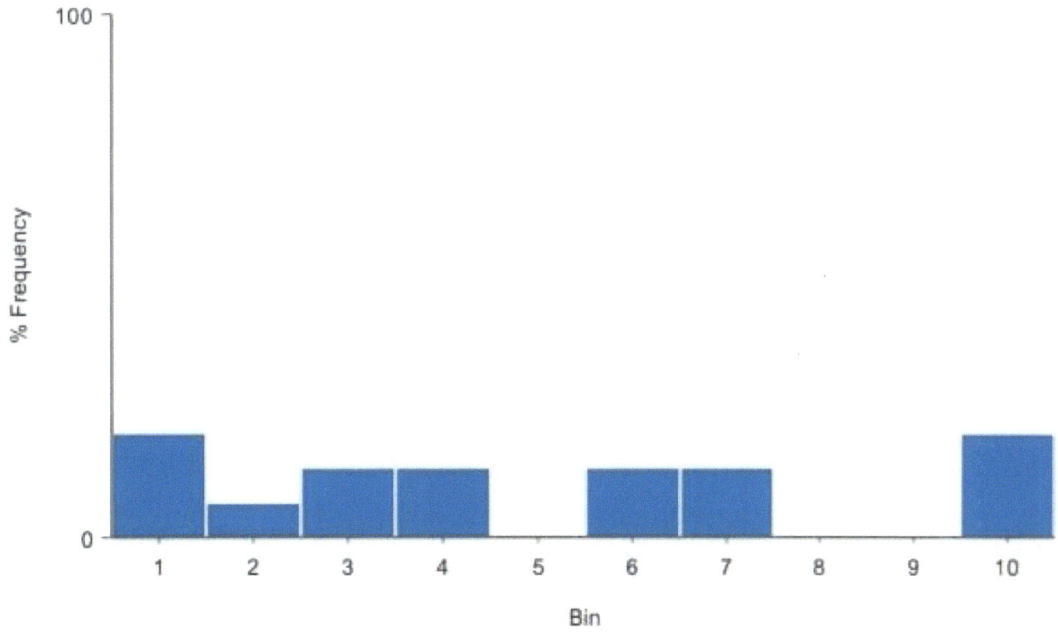

Figure 8: UTS (Mpa) distribution

Box plots display information about the minimum, maximum, first quartile, second quartile or median and third quartile of every variable in the data set. They consist of two parts: a box and two whiskers. The length of the box represents the interquartile range (IQR), which is the distance between the third quartile and the first quartile. The middle half of the data falls inside the interquartile range. The whisker below the box shows the minimum of the variable while the whisker above the box shows the maximum of the variable. Within the box, it will also be drawn a line which represents the median of the variable. Box plots also provide information about the shape of the data. If most of the data are concentrated between the median and the maximum, the distribution is skewed right, if most of the data are concentrated between the median and the minimum, it is said that the distribution is skewed left and if there is the same number of values at the both sides of the median, the distribution is said to be symmetric. The following chart shows the box plot for the variable Tool Rotational Speed (RPM). The minimum of the variable is 400, the first quartile is 400, the second quartile or median is 600, the third quartile is 800 and the maximum is 1000.

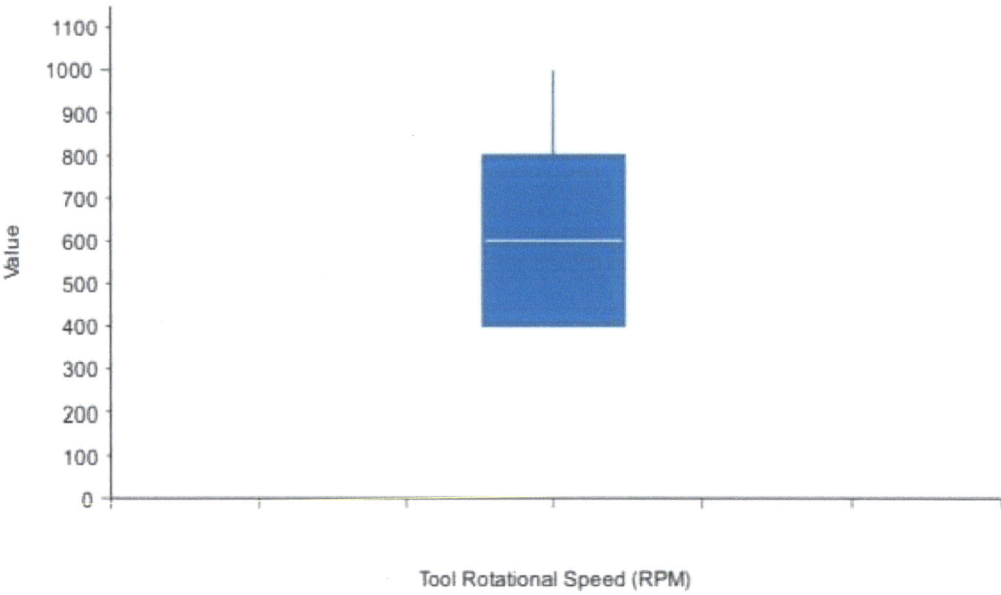

Figure 9: Tool Rotational Speed (RPM) box plot

The following chart shows the box plot for the variable Welding Speed (mm/min). The minimum of the variable is 10, the first quartile is 10, the second quartile or median is 20, the third quartile is 30 and the maximum is 40.

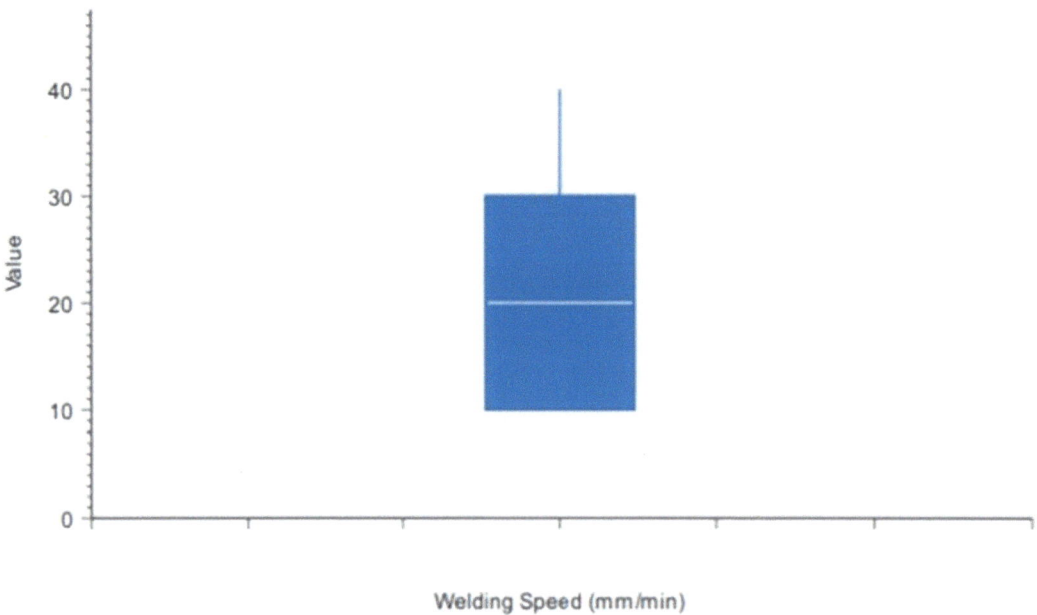

Figure 10: Welding Speed (mm/min) box plot

The following chart shows the box plot for the variable Elongation (%). The minimum of the variable is 2.73, the first quartile is 3.27, the second quartile or median is 3.85, the third quartile is 5.13 and the maximum is 5.92.

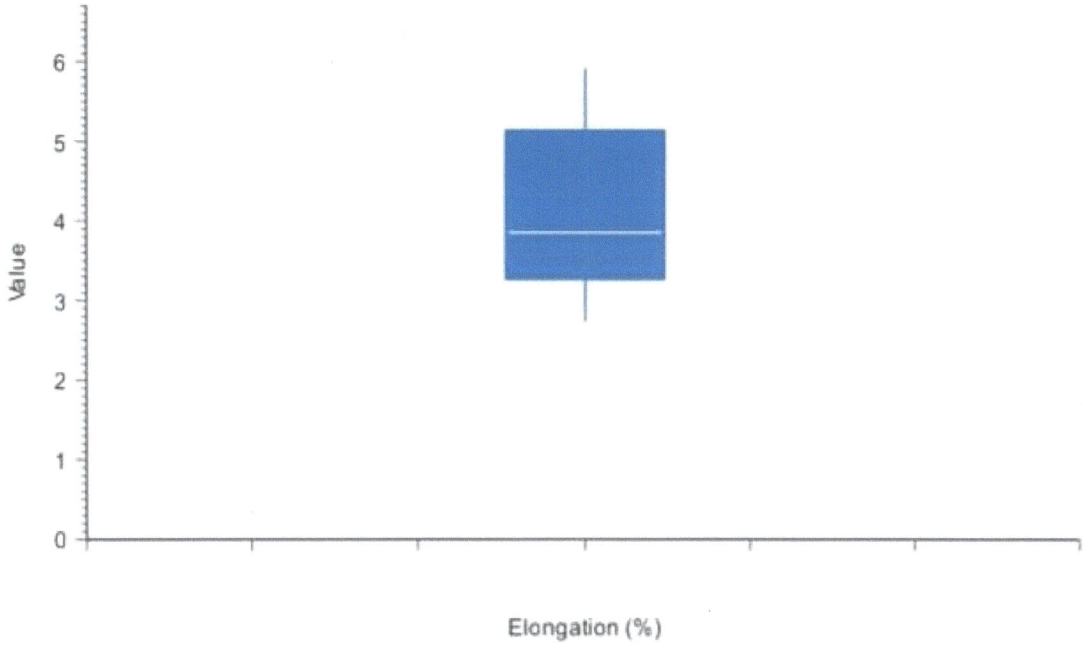

Figure 11: Elongation % box plot

The following chart shows the box plot for the variable UTS (MPa). The minimum of the variable is 138.16, the first quartile is 145.51, the second quartile or median is 161.14, the third quartile is 184.1 and the maximum is 204.25.

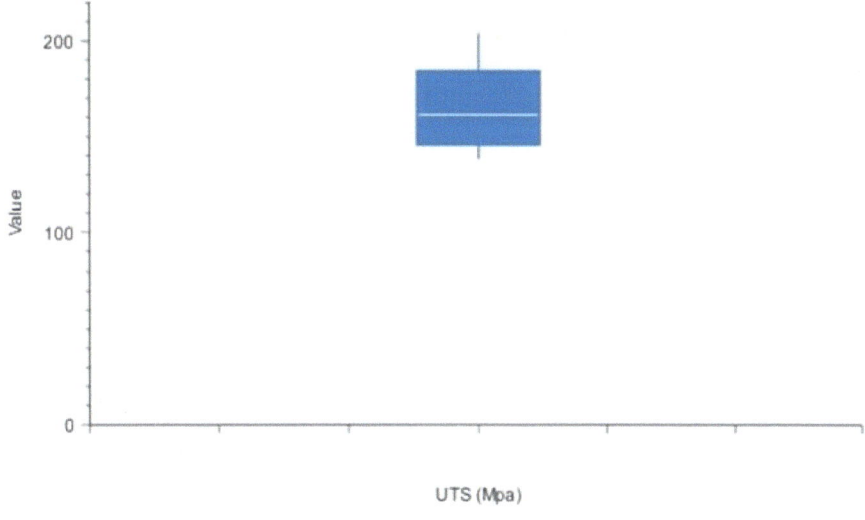

Figure 12: UTS (MPa) box plot

Target balancing task balances the distribution of targets in a data set for function regression. It unuses a given percentage of the instances whose values belong to the most populated bins. After this process, the distribution of the data will be more uniform and, in consequence, the resulting model will probably be of better quality. The percentage of unused instances has been 10%, which corresponds to 1 instances. The following chart shows the histogram for the target variable Elongation (%). The abscissa represents the centers of the containers, and the ordinate their corresponding frequencies. The minimum frequency is 0, which corresponds to the bins with centers 4.73 and 5.34. The maximum frequency is 2, which corresponds to the bins with centers 2.88, 3.19, 3.5, 3.8, 4.11 and 5.65.

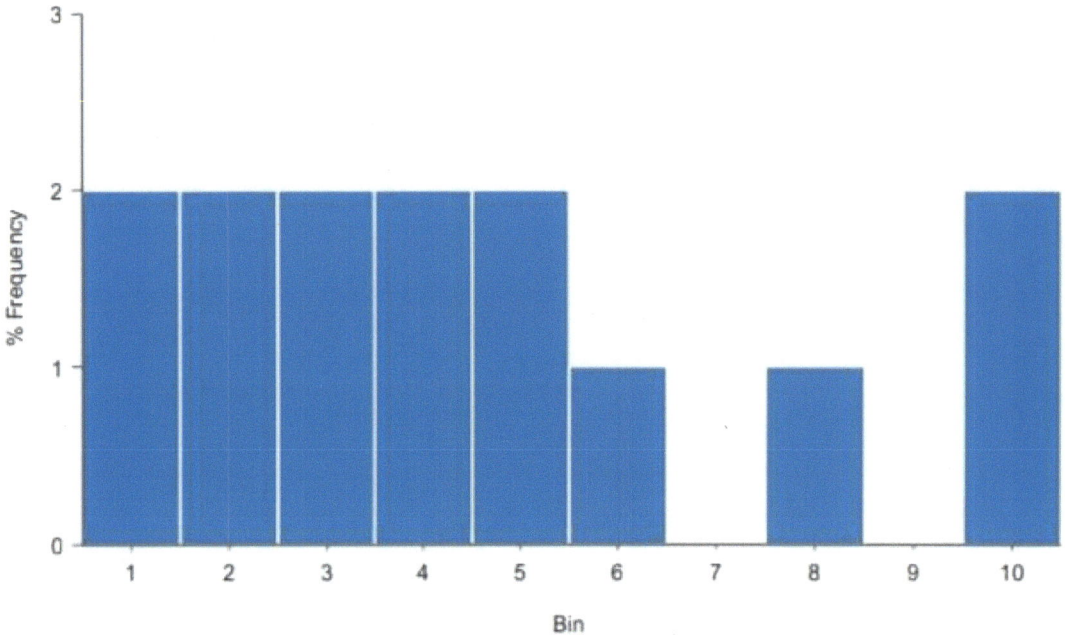

Figure 13: Elongation (%) histogram

The percentage of unused instances has been 10%, which corresponds to 1 instances. The following chart shows the histogram for the target variable UTS (Mpa). The abscissa represents the centers of the containers, and the ordinate their corresponding frequencies. The minimum frequency is 0, which corresponds to the bins with centers 168, 188 and 194. The maximum frequency is 3, which corresponds to the bins with centers 141 and 201.

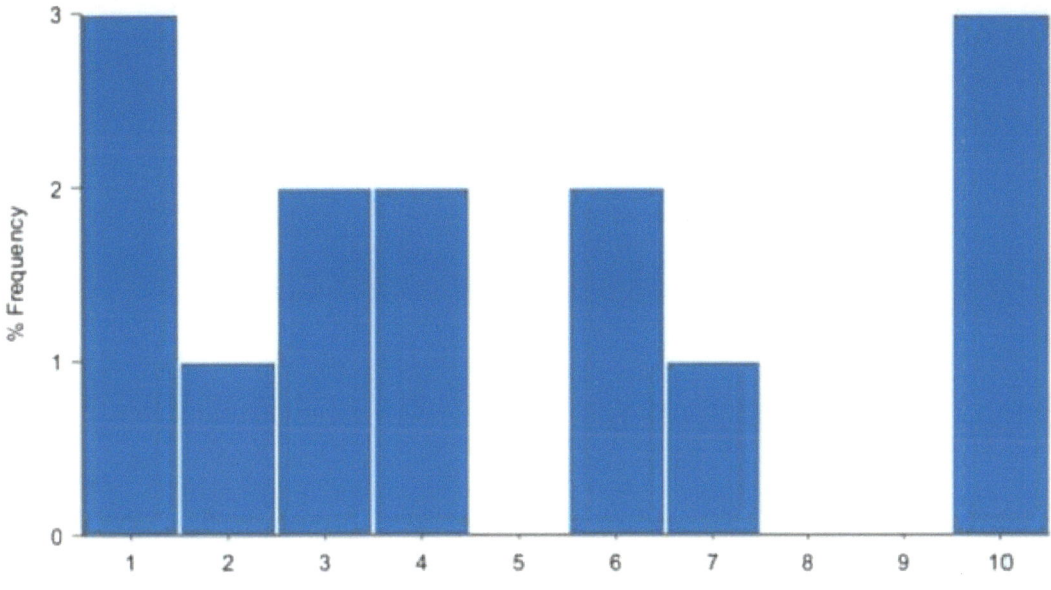

Figure 14: UTS (MPa) histogram

Scatter plot task plots graphs of all targets versus all input variables. That charts might help to see the dependencies of the targets with the inputs. The following chart shows the scatter plot for the input Tool Rotational Speed (RPM) and the target Elongation (%).

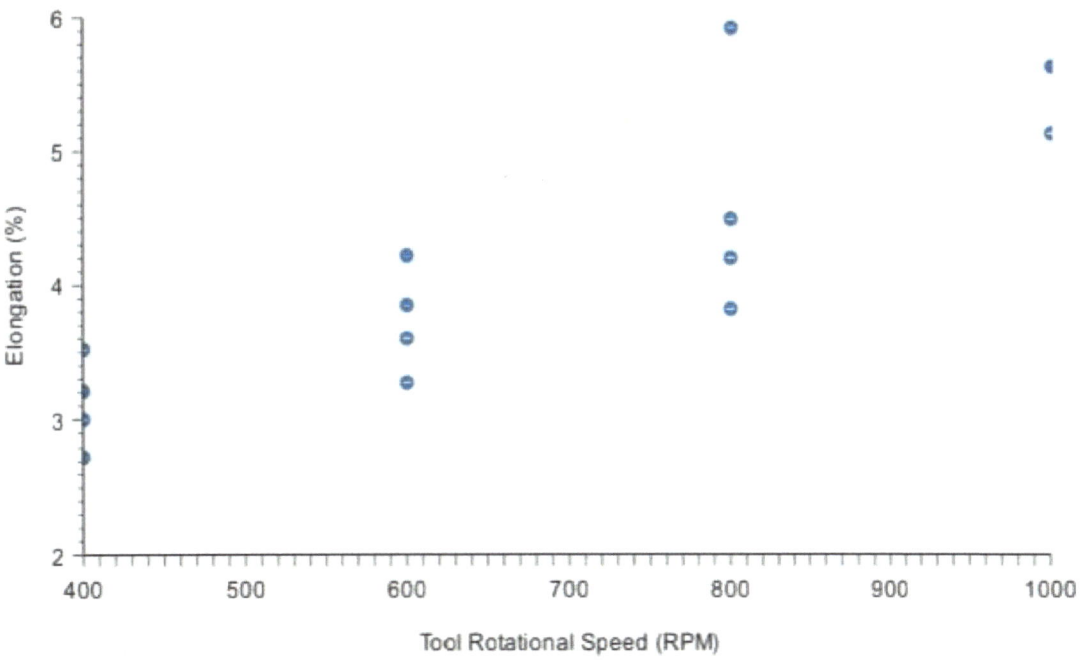

Figure 15: Elongation (%) scatter chart vs Tool Rotational Speed (RPM)

The following chart shows the scatter plot for the input Tool Rotational Speed (RPM) and the target UTS (MPa).

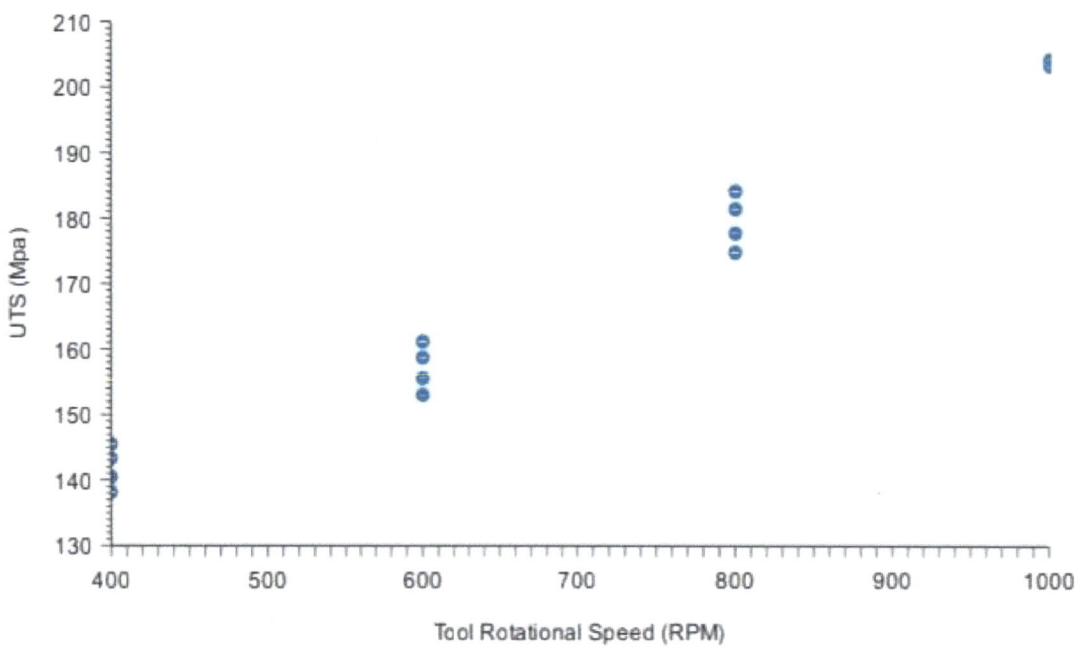

Figure 16: UTS (MPa) scatter chart vs Tool Rotational Speed (RPM)

The following chart shows the scatter plot for the input Welding Speed (mm/min) and the target Elongation (%).

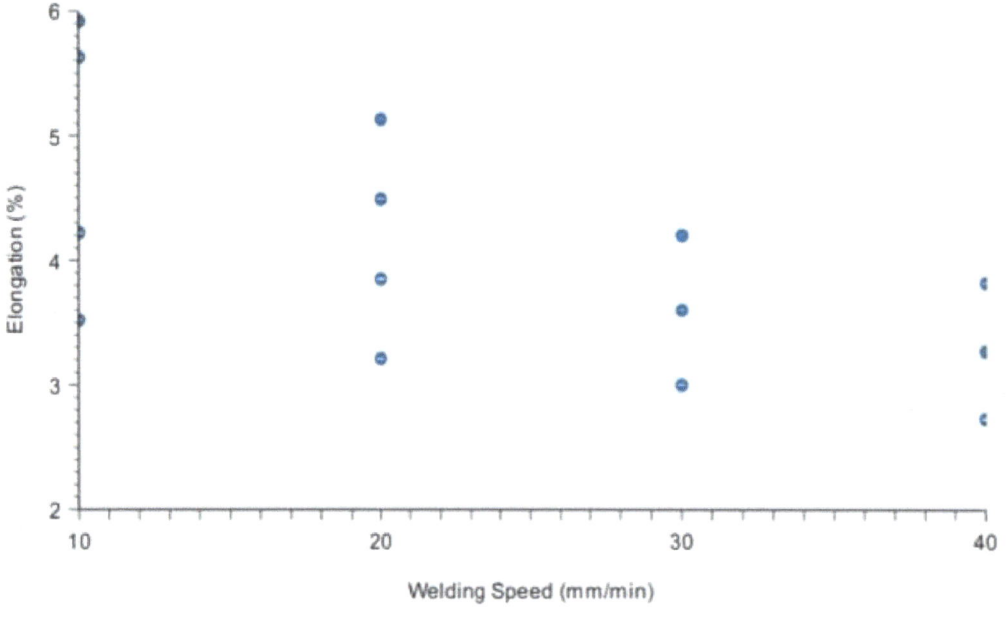

Figure 17: Elongation (%) scatter chart vs Welding Speed (mm/min)

The following chart shows the scatter plot for the input Welding Speed (mm/min) and the target UTS (MPa).

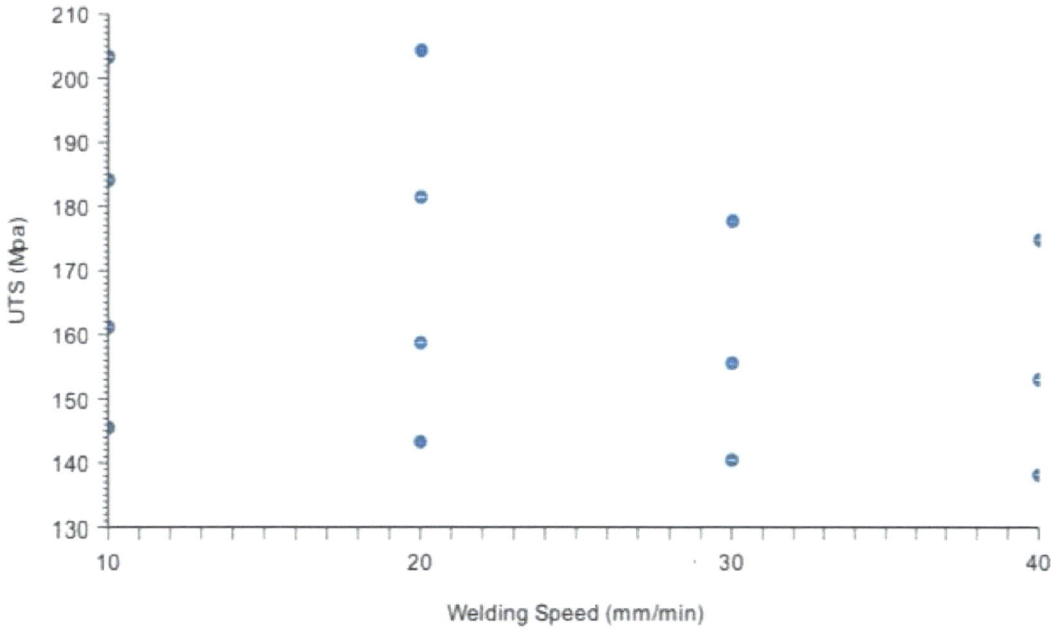

Figure 18: UTS (MPa) scatter chart vs Welding Speed (mm/min)

Correlation matrix task calculates the absolute values of the linear correlations among all inputs. The correlation is a numerical value between 0 and 1 that expresses the strength of the relationship between two variables. When it is close to 1 it indicates a strong relationship, and a value close to 0 indicates that there is no relationship. The following table shows the absolute value of the correlations between all input variables. The minimal correlation is 0.0918292 between the variables Tool Rotational Speed (RPM) and Welding Speed (mm/min). The maximal correlation is 0.0918292 between the variables Tool Rotational Speed (RPM) and Welding Speed (mm/min).

Table 5: Correlation matrix

	Tool Rotational Speed (RPM)	Welding Speed (mm/min)
Tool Rotational Speed (RPM)	1	0.0918
Welding Speed (mm/min)		1

It might be interesting to look for linear dependencies between single input and single target variables. This task calculates the absolute values of the correlation coefficient between all inputs and all targets. Correlations close to 1 mean that a single target is linearly correlated with a single input. Correlations close to 0 mean that there is not a linear relationship between an input and a target variables. Note that, in general, the targets depend on many inputs simultaneously. The following table shows the absolute value of the linear correlations between all input and target variables. The maximum correlation (0.986308) is yield between the input variable Tool Rotational Speed (RPM) and the target variable UTS (MPa).

Table 6: Elongation (%) linear correlations

	Elongation (%)
Tool Rotational Speed (RPM)	0.869
Welding Speed (mm/min)	-0.492

The next chart illustrates the dependency of the target Elongation (%) with all the input variables.

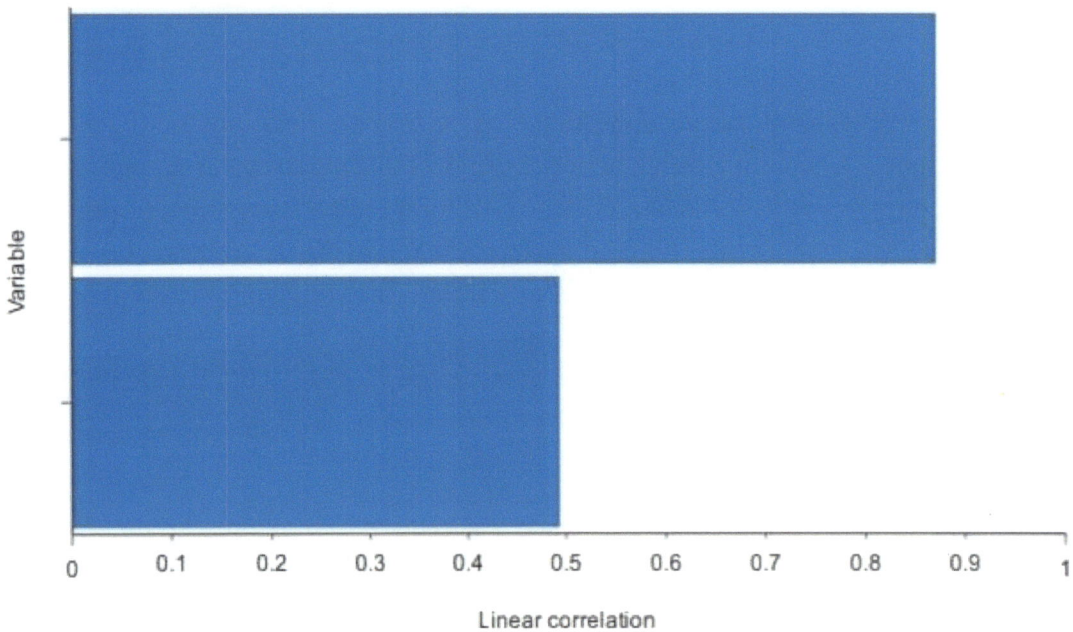

Figure 19: Elongation (%) bars chart

The following table shows the absolute value of the linear correlations between all input and target variables. The maximum correlation (0.986308) is yield between the input variable Tool Rotational Speed (RPM) and the target variable UTS (Mpa).

Table 7: UTS (MPa) linear correlations

	UTS (Mpa)
Tool Rotational Speed (RPM)	0.986
Welding Speed (mm/min)	-0.273

The next chart illustrates the dependency of the target UTS (MPa) with all the input variables.

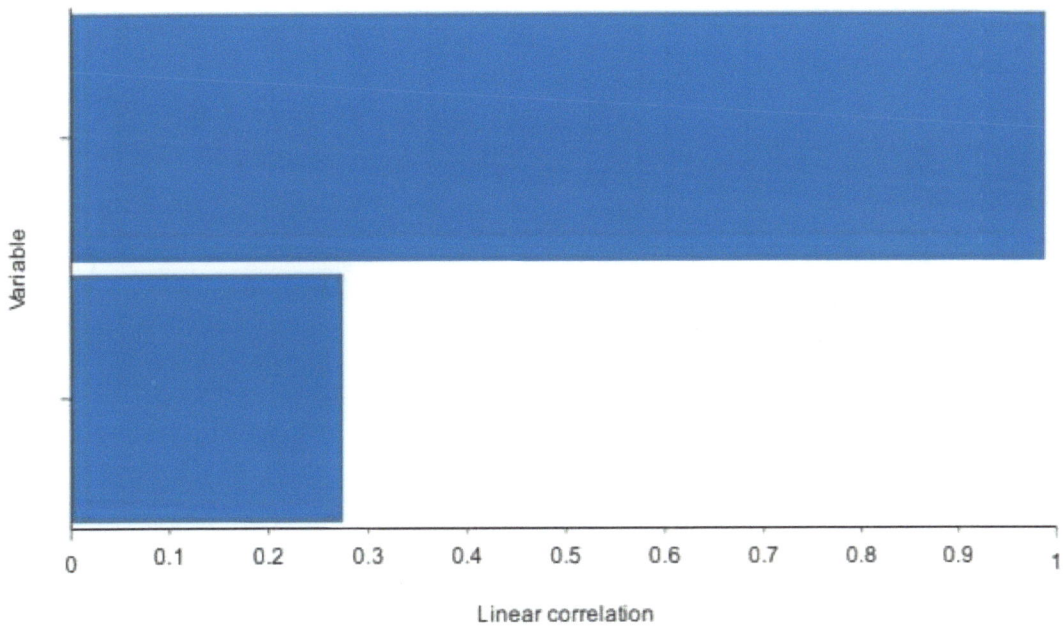

Figure 20: UTS (MPa) bars chart

Constant variables are those columns in the data matrix having always the same value. They do not provide any information to the model but increase its complexity. Constant variables should neither be used as inputs nor as targets, except when the model will need to include them in the future. There are not constant variables in the data set. Repeated instances are those rows in the data matrix having the same values as other rows. They provide redundant information to the model and should not be used for training, selection or testing. There are no repeated instances in the data set. Outliers are defined as observations in the data that are abnormally distant from the others. They may be due to variability in the measurement or may indicate experimental errors. This task uses the Tukey's method, which defines an outlier as those values of the data set that fall to far from the central point, the median. The maximum distance to the center of the data that is going to be allowed is defined by the cleaning parameter. As it grows, the test becomes less sensitive to outliers but if it is too small, a lot of values will be detected as outliers. The data has not outliers. Outliers are defined as observations in the data that are abnormally distant from the others. They may be due to variability in the measurement or may indicate experimental errors. This task trains a neural network using all the instances, then set as unused the instances that have an error above a given value. The data has not outliers. There are not instances to be filtered in the data set. When designing a predictive model, the general practice is to first divide the data into three subsets. The first subset is the training set, which is used for constructing different candidate models. The second subset is the selection set, which is used to select the model exhibiting the best properties. The third subset is the testing set, which it is used for validating the final model. The following table shows the uses of all the instances in the data set. Note that the instances are arranged in rows of 10. The total number of instances is 15. The numbers of training, selection, testing and unused instances are 10, 2, 2 and 1, respectively.

Table 8: Instances table

	1	2	3	4	5	6	7	8	9	10
0	Train.	Train.	Test.	Test.	Train.	Train.	Train.	Train.	Unused	Train.
10	Train.	Sel.	Train.	Sel.	Train.					

The following pie chart details the uses of all the instances in the data set. There are 10 instances for training (66.7%), 2 instances for selection (13.3%), 2 instances for testing (13.3%) and 1 unused instances (6.67%).

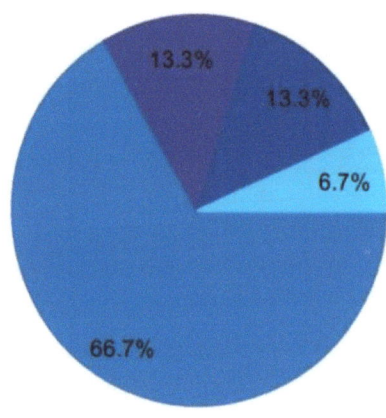

Figure 21: Instances Pie chart

Principal components analysis allows to identify underlying patterns in a data set so it can be expressed in terms of other data set of lower dimension without much loss of information. The resulting data set should be able to explain most of the variance of the original data set by making a variable reduction. The final variables will be named principal components. Since this process is not reversible, it will be only applied to the input variables.

The next chart shows the cumulative explained variance for the principal components. The x-axis represents each of the principal components and the y-axis depicts the cumulative explained variance. As it can be seen, the total explained variance for all the principal components is 100% but if the number of chosen principal components decreases also makes it the total explained variance.

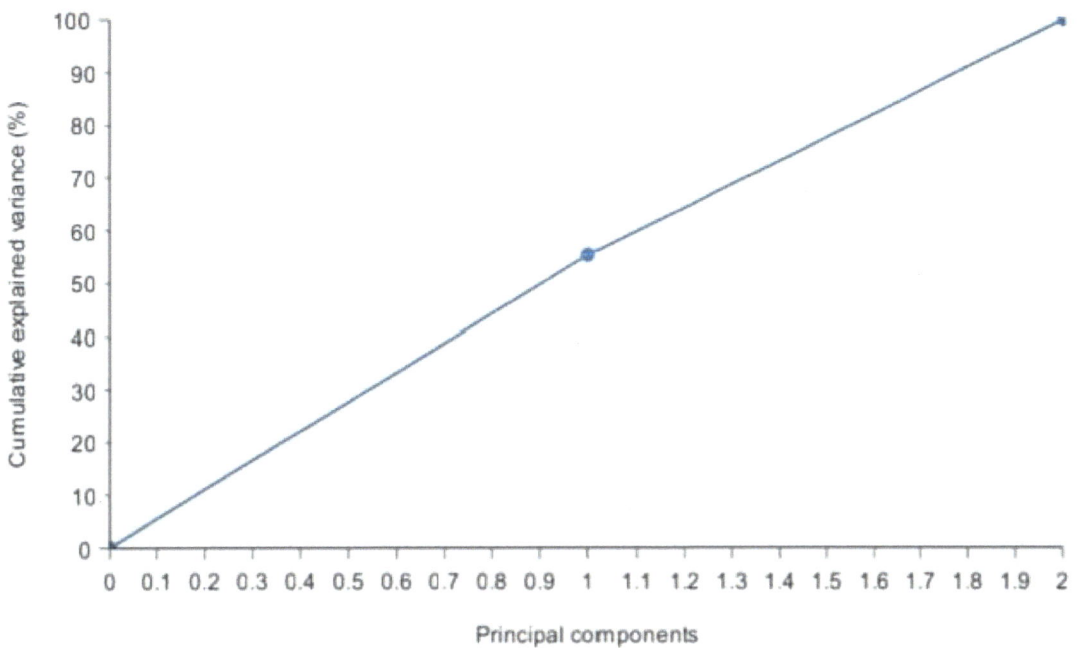

Figure 22: Explained variance chart

3.2 Neural Network

The neural networks represent the predictive model. In Neural Designer neural networks allow deep architectures, which are a class of universal approximator. The size of the scaling layer is 2, the number of inputs. The scaling method for this layer is the Minimum-Maximum. The following table shows the values which are used for scaling the inputs, which include the minimum, maximum, mean and standard deviation.

Table 9: Scaling layer

	Minimum	Maximum	Mean	Deviation
Tool Rotational Speed (RPM)	400	1e+003	680	224
Welding Speed (mm/min)	10	40	24	11.2

The number of layers in the neural network is 2. The following table depicts the size of each layer and its corresponding activation function. The architecture of this neural network can be written as 2:3:2.

Table 10: Number of layers and Activation function

	Inputs number	Neurons number	Activation function
1	2	3	Hyperbolic Tangent
2	3	2	Linear

The following table shows the statistics of the parameters of the neural network. The total number of parameters is 17.

Table 11: Neural Network parameters

	Minimum	Maximum	Mean	Standard deviation
Statistics	-2.01	2.31	0.0876	1.1

The size of the unscaling layer is 2, the number of outputs. The unscaling method for this layer is the minimum and maximum. The following table shows the values which are used for scaling the inputs, which include the minimum, maximum, mean and standard deviation.

Table 12: Unscaling layer

	Minimum	Maximum	Mean	Deviation
Elongation (%)	2.73	5.8	4.03	0.949
UTS (Mpa)	138	204	167	23.3

A graphical representation of the network architecture is depicted next. It contains a scaling layer, a neural network and an unscaling layer. The yellow circles represent scaling neurons, the green circles the principal components, the blue circles perceptron neurons and the red circles unscaling neurons. The number of inputs is 2, the number of principal components is 2, and the number of outputs is 2. The complexity, represented by the numbers of hidden neurons, is 3.

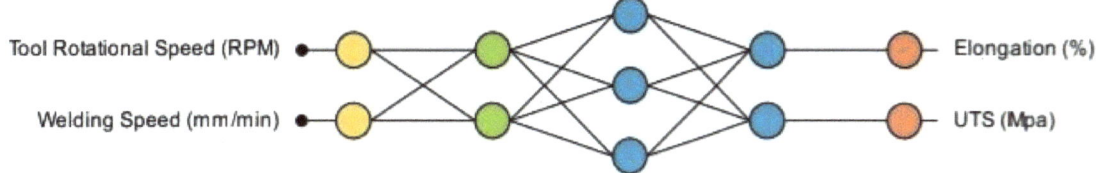

Figure 23: Neural Network Architecture

The loss index plays an important role in the use of a neural network. It defines the task the neural network is required to do, and provides a measure of the quality of the representation that it is required to learn. The choice of a suitable loss index depends on the particular application. The normalized squared error is used here as the error method. It divides the squared error between the outputs from the neural network and the targets in the data set by a normalization coefficient. If the normalized squared error has a value of unity then the neural network is predicting the data 'in the mean', while a value of zero means perfect prediction of the data. The neural parameters norm is used as the regularization method. It is applied to control the complexity of the neural network by reducing the value of the parameters. The procedure used to carry out the learning process is called training (or learning) strategy. The training strategy is applied to the neural network in order to obtain the best possible loss. The quasi-Newton method is used here as training algorithm. It is based on Newton's method, but does not require calculation of second derivatives. Instead, the quasi-Newton method computes an approximation of the inverse Hessian at each iteration of the algorithm, by only using gradient information. The following plot shows the losses in each iteration. The initial value of the training loss is 1.09861, and the final value after 87 iterations is 0.109506. The initial value of the selection loss is 67.587, and the final value after 87 iterations is 154.445.

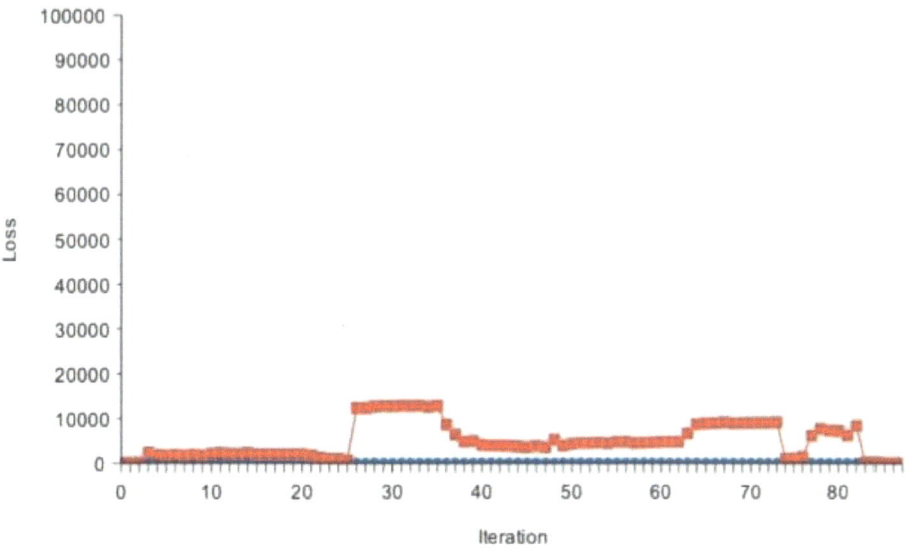

Figure 24: Quasi-Newton method losses history

The next table shows the training results by the quasi-Newton method. They include some final states from the neural network, the loss functional and the training algorithm.

Table 13: Quasi-Newton method results

	Value
Final parameters norm	106
Final loss	0.11
Final selection loss	154
Final gradient norm	0.000868
Iterations number	87
Elapsed time	00:01
Stopping criterion	Gradient norm goal

3.3 Model Selection

Model selection is applied to find a neural network with a topology that optimizes the loss on new data. There are two different types of algorithms for model selection: Order selection algorithms and input selection algorithms. Order selection algorithms are used to find the optimal number of hidden neurons in the network. Inputs selection algorithms are responsible for finding the optimal subset of input variables. The inputs selection algorithm chosen for this application is growing inputs. With this method, the inputs are added progressively based on their correlations with the targets. The order selection algorithm chosen for this application is incremental order. This method start with the minimum order and adds a given number of perceptrons in each iteration.

Input importance task calculates the selection loss when removing one input at a time. This shows which input have more influence in the outputs. The next table shows the importance of each input. If the importance takes a value greater than 1 for an input, it means that the selection error without that input is greater than with it. In the case that the importance is lower than 1, the selection error is lower without using that input. Finally, if the importance is 1, there is no difference between using the current input and not using it. The most important variable is Welding Speed (mm/min), that gets a contribution of 123.3% to the outputs.

Table 14: Inputs importance results

	Contribution
Tool Rotational Speed (RPM)	0.9
Welding Speed (mm/min)	1.23

The next chart illustrates the contribution of each input.

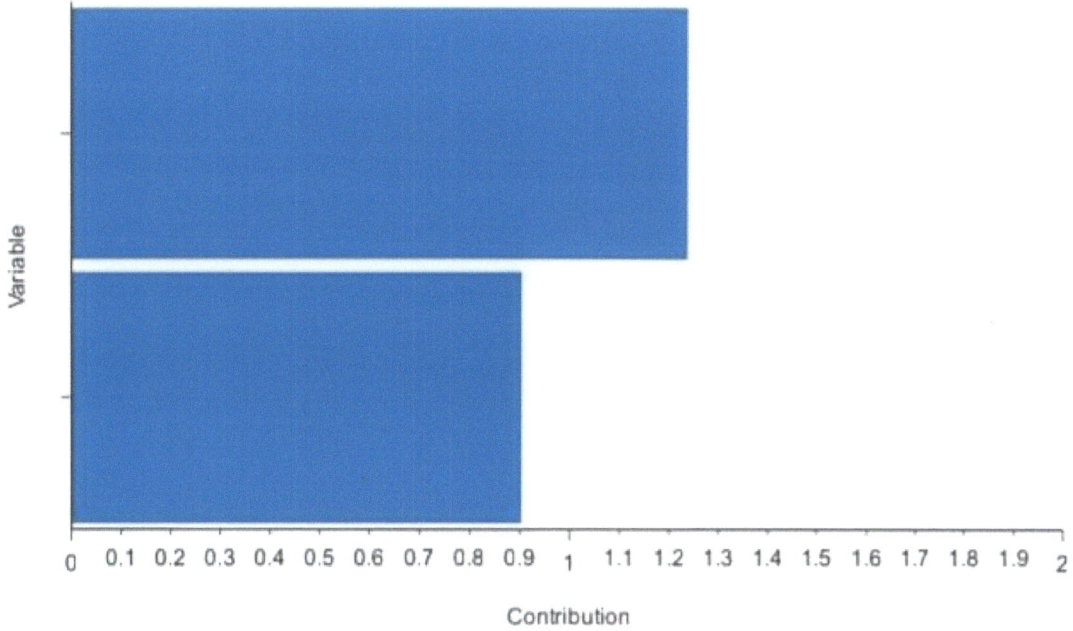

Figure 25: Contribution bars chart

The best selection is achieved by using a model whose complexity is the most appropriate to produce an adequate fit of the data. The order selection algorithm is responsible of finding the optimal number of neurons in the network. Incremental order is used here as order selection algorithm in the model selection. The next chart shows the loss history for the different subsets during the incremental order selection process. The blue line represents the training loss and the red line symbolizes the selection loss.

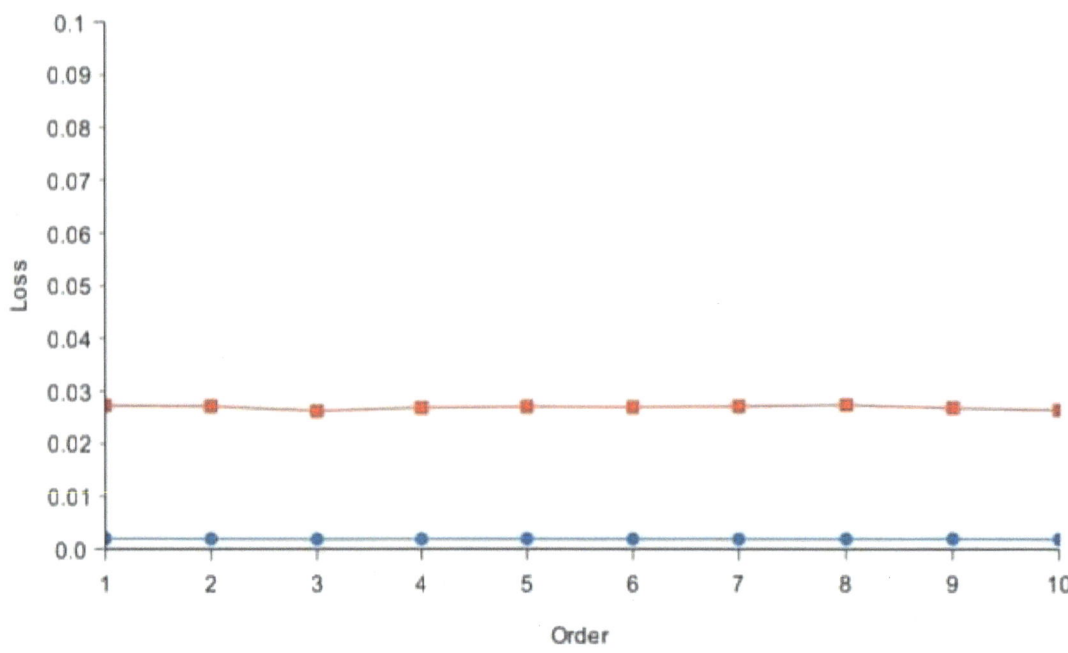

Figure 26: Incremental order losses plot

The next table shows the order selection results by the incremental order algorithm. They include some final states from the neural network, the loss functional and the order selection algorithm.

Table 15: Incremental order results

	Value
Optimal order	1
Optimum training loss	0.00202364
Optimum selection loss	0.0272754
Iterations number	10
Elapsed time	00:10

A graphical representation of the resulted deep architecture is depicted next. It contains a scaling layer, a neural network and an unscaling layer. The yellow circles represent scaling neurons, the green circles the principal components, the blue circles perceptron neurons and the red circles unscaling neurons. The number of inputs is 2, the number of principal components is 2, and the number of outputs is 2. The complexity, represented by the numbers of hidden neurons, is 1.

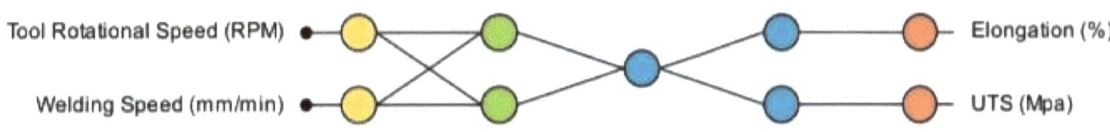

Figure 27: Final architecture

A standard method to test the loss of a model is to perform a linear regression analysis between the scaled neural network outputs and the corresponding targets for an independent testing subset. This analysis leads to 3 parameters for each output variable. The first two parameters, a and b, correspond to the y-intercept and the slope of the best linear regression relating scaled outputs and targets. The third parameter, R2, is the correlation coefficient between the scaled outputs and the targets. If we had a perfect fit (outputs exactly equal to targets), the slope would be 1, and the y-intercept would be 0. If the correlation coefficient is equal to 1, then there is a perfect correlation between the outputs from the neural network and the targets in the testing subset. The next plot lists the linear regression parameters for the scaled output Elongation (%).

Table 16: Elongation (%) linear regression parameters

	Value
Intercept	-0.299
Slope	0.698
Correlation	1

The next chart illustrates the linear regression for the scaled output Elongation (%). The predicted values are plotted versus the actual ones as squares. The coloured line indicates the best linear fit. The grey line would indicate a perfect fit.

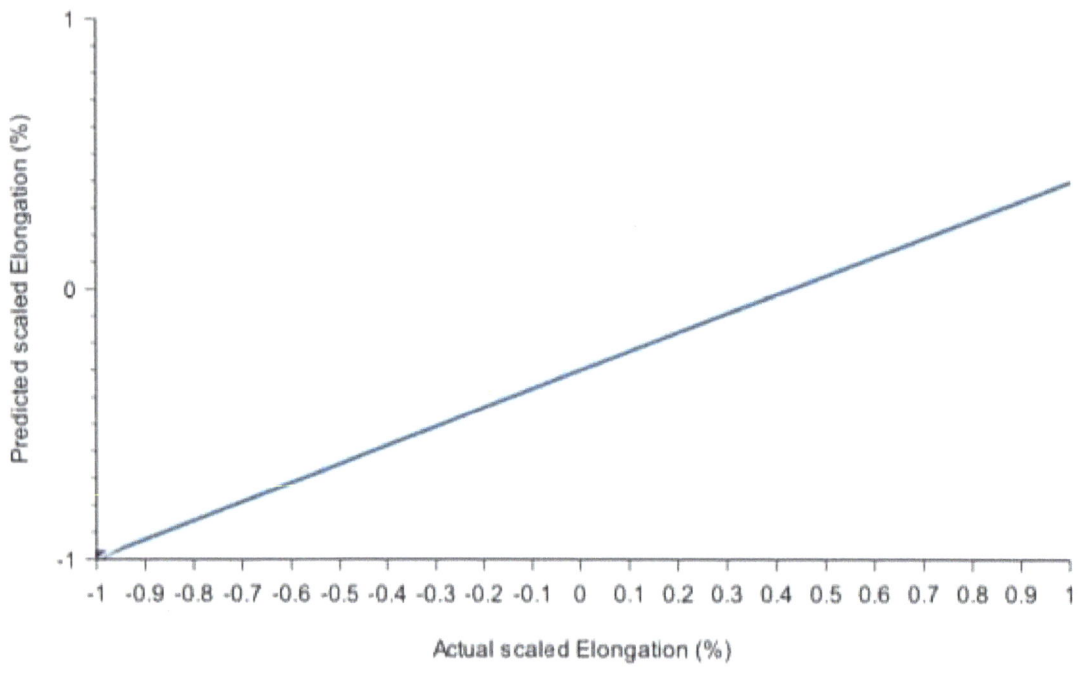

Figure 28: Elongation (%) linear regression chart

The next plot lists the linear regression parameters for the scaled output UTS (Mpa).

Table 17: UTS (MPa) linear regression parameters

	Value
Intercept	0.986
Slope	1.98
Correlation	1

The next chart illustrates the linear regression for the scaled output UTS (Mpa). The predicted values are plotted versus the actual ones as squares. The coloured line indicates the best linear fit. The grey line would indicate a perfect fit.

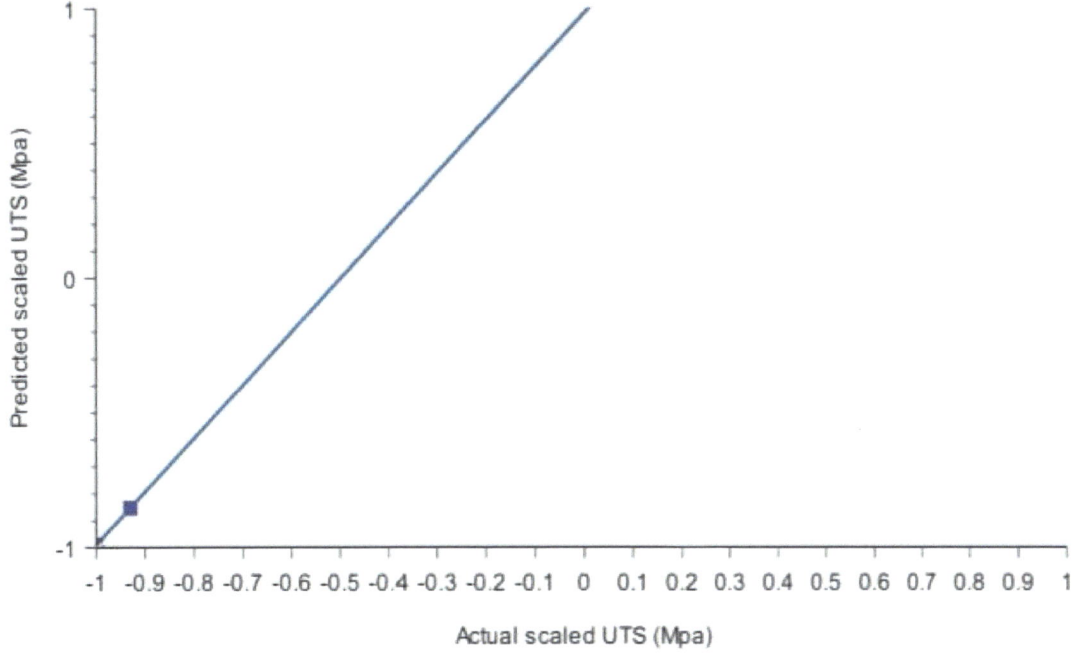

Figure 29: UTS (MPa) linear regression chart

Testing errors task measures all the losses of the model. it takes in account every used instance and evaluate the model for each use. The next table shows all the errors of the data for each use of them.

Table 18: Errors table

	Training	Selection	Testing
Sum squared error	1.43447	0.165925	0.0855475
Mean squared error	0.143447	0.0829623	0.0427738
Root mean squared error	0.378744	0.288032	0.206818
Normalized squared error	0.26513	0.83134	2.48148
Minkowski error	2.0285	0.281676	0.185371

A neural network produces a set of outputs for each set of inputs applied. The outputs depend, in turn, on the values of the parameters. The next table shows the input values and their corresponding output values. The input variables are Tool Rotational Speed (RPM) and Welding Speed (mm/min); and the output variables are Elongation (%) and UTS (MPa).

Table 19: Inputs-outputs table

	Value
Tool Rotational Speed (RPM)	1000
Welding Speed (mm/min)	40
Elongation (%)	4.42639577
UTS (Mpa)	173.530702

It is very useful to see the how the outputs vary as a function of a single input, when all the others are fixed. This can be seen as the cut of the neural network model along some input direction and through some reference point. The next plot shows the output Elongation (%) as a function of the input Tool Rotational Speed (RPM). The x and y axes are defined by the range of the variables Tool Rotational Speed (RPM) and Elongation (%), respectively.

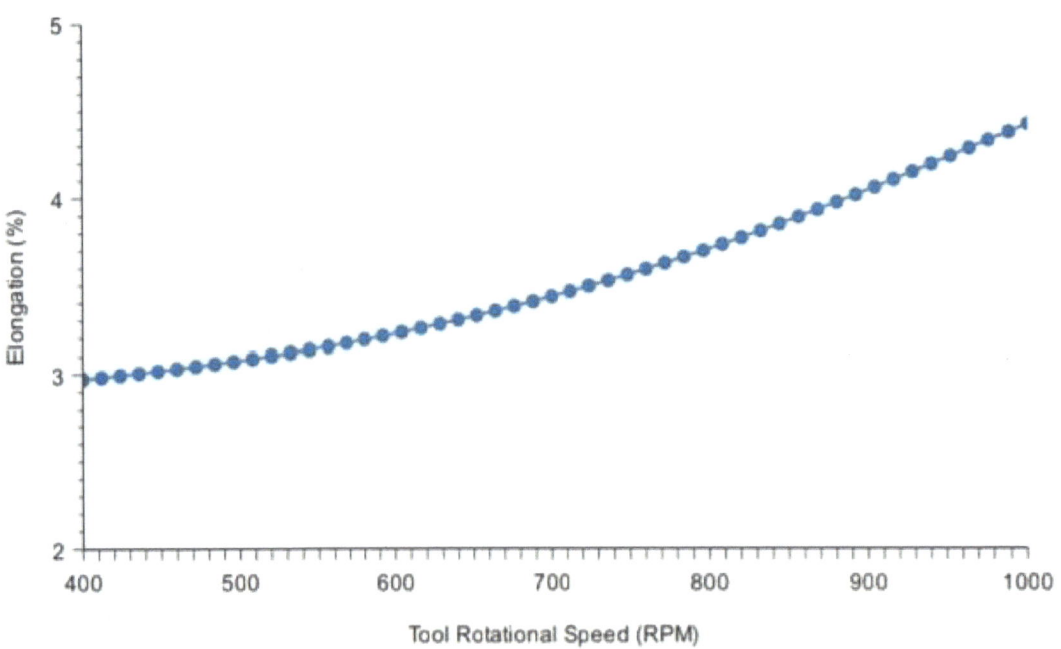

Figure 30: Elongation (%) against Tool Rotational Speed (RPM) directional line chart

The next plot shows the output UTS (MPa) as a function of the input Tool Rotational Speed (RPM). The x and y axes are defined by the range of the variables Tool Rotational Speed (RPM) and UTS (MPa), respectively.

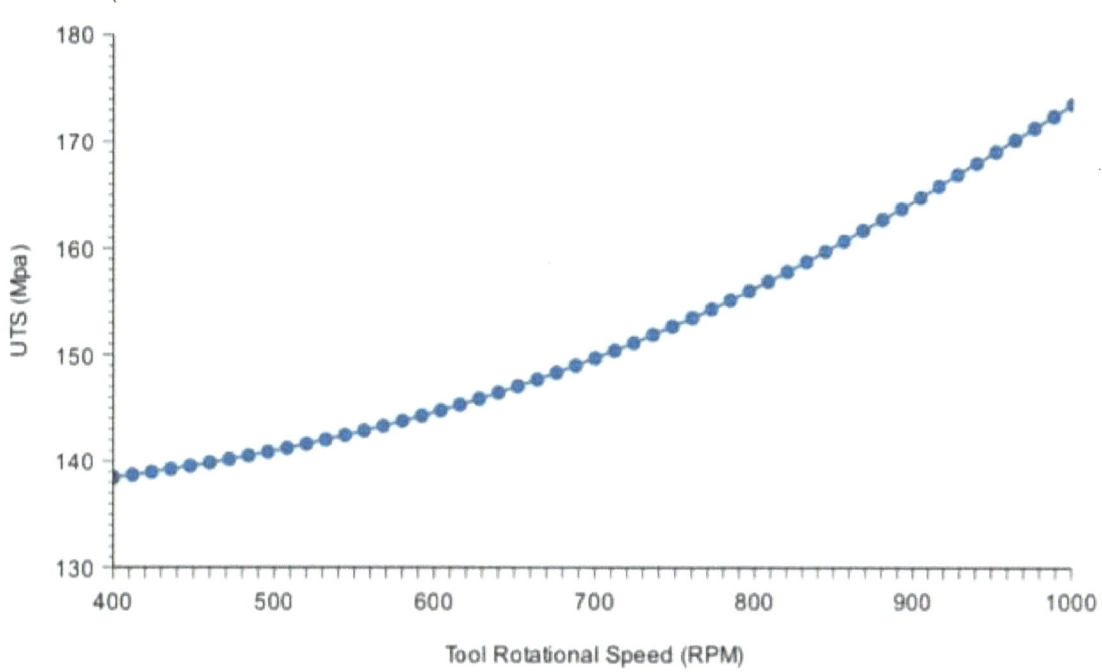

Figure 31: UTS (MPa) against Tool Rotational Speed (RPM) directional line chart

The next plot shows the output Elongation (%) as a function of the input Welding Speed (mm/min). The x and y axes are defined by the range of the variables Welding Speed (mm/min) and Elongation (%), respectively.

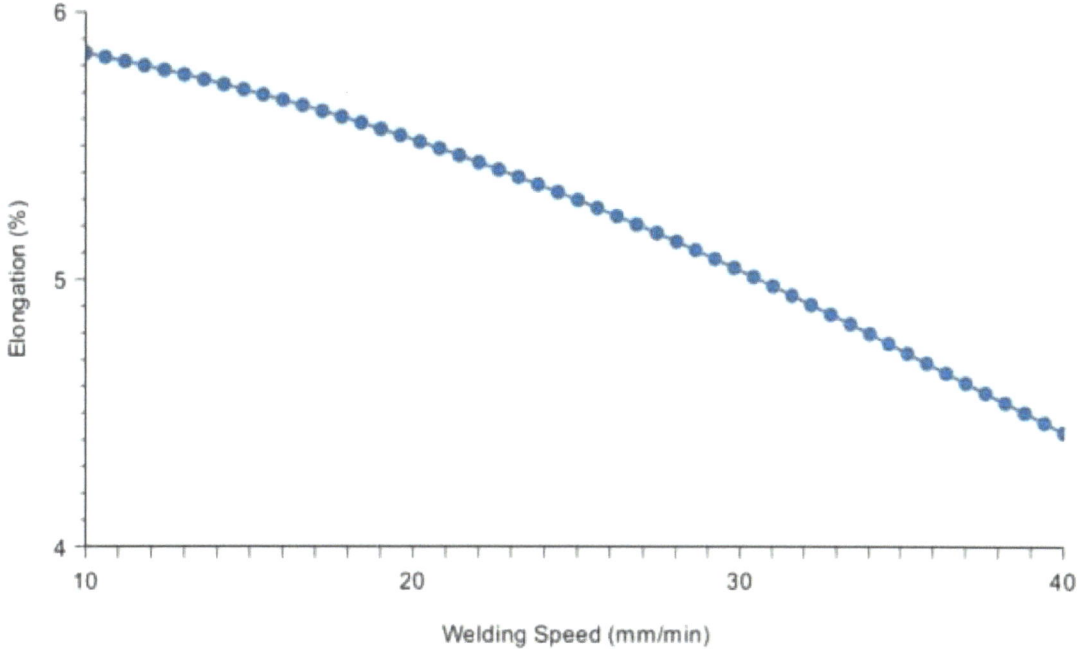

Figure 32: Elongation (%) against Welding Speed (mm/min) directional line chart

The next plot shows the output UTS (Mpa) as a function of the input Welding Speed (mm/min). The x and y axes are defined by the range of the variables Welding Speed (mm/min) and UTS (Mpa), respectively. Note that some directional outputs fall outside the range of UTS (Mpa), and therefore they are not plotted.

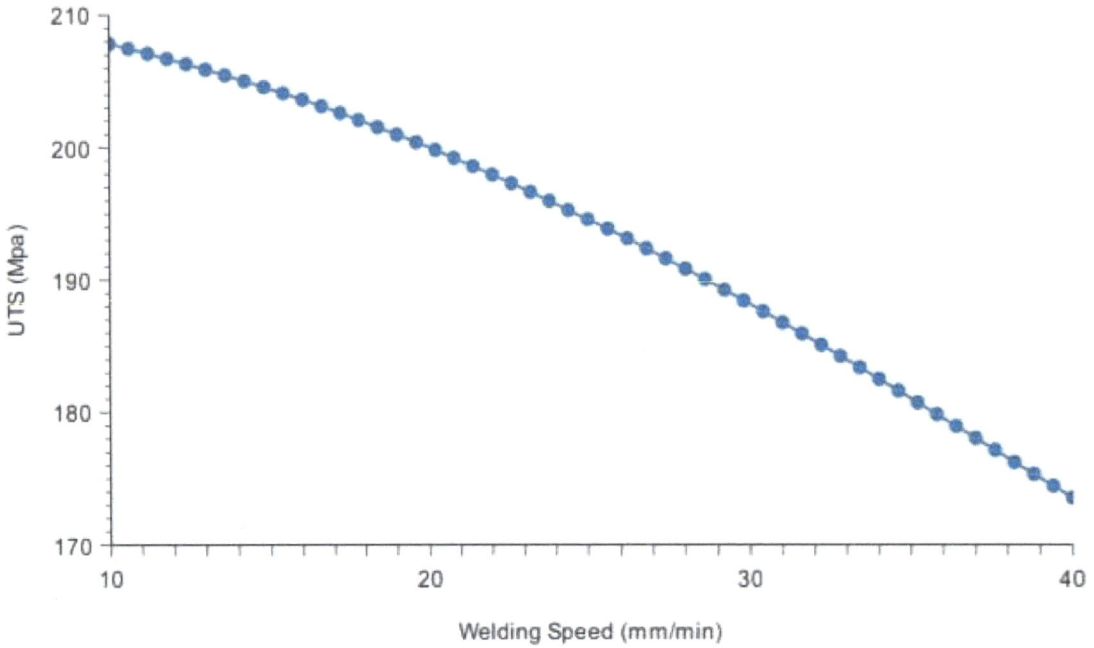

Figure 33: UTS (MPa) against Welding Speed (mm/min) directional line chart

4. Conclusion

Comparison of the predicted and the experimental results confirms that the networks are adjusted carefully, and the ANN can be used for modeling of FSW effective parameters. Accuracy of 98.47% and 80.73% was achieved for predicting the values of the elongation % and Ultimate Tensile Strength (UTS) respectively.

References

1. DUTTA P, PRATIHAR D K. Modeling of TIG welding process using conventional regression analysis and neural network-based approaches [J]. Journal of Materials Processing Technology, 2007, 184: 56í68.

2. ATES H. Prediction of gas metal arc welding parameters based on artificial neural networks [J]. Materials and Design, 2007, 28: 2015í2023.

3. OKUYUCU H, KURT A, ARCAKLIOGLU E. Artificial neural network application to the friction stir welding of aluminum plates [J]. Materials and Design, 2007, 28(1): 78í84.

4. Tansel, I. N., Demetgul, M., Okuyucu, H., & Yapici, A. (2010). Optimizations of friction stir welding of aluminum alloy by using genetically optimized neural network. The International Journal of Advanced Manufacturing Technology, 48(1-4), 95-101.

Prediction of the mechanical properties of Friction Stir Welded Joints of aerospace alloys using Artificial Neural Network

Akshansh Mishra[1], Dr. A. Razal Rose[2], Katyayani Jaiswal[3]

[1]Project Scientific Officer, Stir Research Technologies, Uttar Pradesh, India
[2]Faculty of Mechanical Engineering, SRM Institute of Science and Technology, Kattangulathur
[3]Department of Computer Science, IIT Ropar, Punjab

Abstract: Artificial Neural Networks (ANNs) have a built in capability to adapt their synaptic weights to changes in surroundings in the environment. Artificial Neural networks find application in areas of prediction and classification, the areas where statistical methods have traditionally been used. In our present work we designed Neural Network for predicting the tensile strength of Friction Stir Welded dissimilar aerospace alloys joints. Quasi Newton algorithm is used for training the Neural Networks. Tool rotational Speed (rpm), traverse speed (mm/min) were the inputs while Vickers Hardness (HV) and Tensile Strength (MPa) were outputs in the Neural Network architecture. The accuracy obtained between the experimental values and predicted values of the tensile strength and the Vickers hardness is 98.07% and 83.59%.

Keywords: Friction Stir Welding; Neural Network; Quasi Newton Algorithm; Mechanical Properties.

1. Introduction:

Artificial Neural Network (ANN) is a massively parallel distributed processor made up of simple processing units that has a natural propensity for storing experimental knowledge and making it available for use. It resembles the brain in two respects:

a) Knowledge is acquired by the network from its environment through a learning process.

b) Interneuron connection strengths known as synaptic weights are used to store the acquired knowledge.

The procedure used to perform the learning process is called a learning algorithm. The function of a learning algorithm is to modify the synaptic weights of the network in an orderly fashion to attain a desired design objective. It is possible for a neural network to modify its own topology which is motivated by the fact that neurons in the human brain can die and new synaptic neurons can grow. A neuron is an information processing unit that is fundamental to the operation of neural network. Figure 1 shows the model of a neuron, which forms the basis for designing a large family of neural networks. Synapses, Adder and an Activation Function are three basic elements of the neural model.

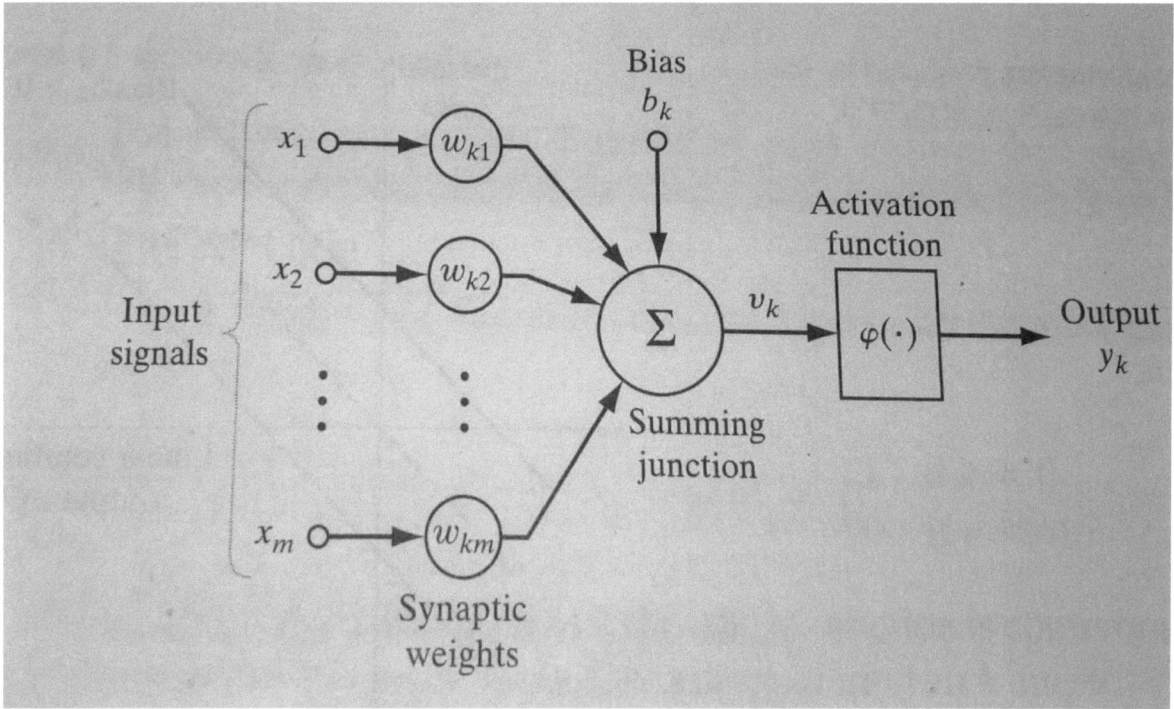

Figure 1: Non- linear model of a Neural Network

A set of synapses or connecting links each of which is characterized by a weight or strength of its own. Specifically, a signal x_j at the input of synapse j connected to neuron k is multiplied by the synaptic weight w_{kj}. It is important to make a note of the manner in which the subscripts of the synaptic weight w_{kj} are written. The first subscript in w_{kj} refers to the neuron in question and the second subscript refers to the input end of the synapse to which the weight refers. Unlike the weight of synapse in the brain, the synaptic weight of an artificial neuron may lie in a range that includes negative as well as positive values.

An adder is used for summing the input signals weighted by the respective synaptic strengths of the neuron. An activation function for limiting the amplitude of the output of a neuron. The activation function is also referred to as squashing function, in that it squashes (limits) the permissible amplitude range of the output signal to some finite value. Typically, the normalized amplitude range of the output of a neuron is written as the closed unit interval [0,1],or alternatively, [-1,1]. The neural model also includes an externally applied bias denoted by b_k. The bias b_k has the effect of increasing or lowering the net input of the activation function, depending on whether it is positive or negative respectively.

ANN has offered great help to the computer integrated manufacturing and the future intelligent manufacturing systems. The application of ANN in engineering design has been focused in two directions. The first is focused on the design of the products based on design rules. The second is focused on configuring functionally complex systems built from standard systems or components, e.g. the design of a manufacturing system [1].

ANNs find wide application in Friction Stir Welding research domain also. Shojaeefard *et al* [2] developed a neural network model to find the correlation between the tool parameters (pin and shoulder diameter) and heat-affected zone, thermal, and strain value in the weld zone.

Jayaraman *et al* [3] predicted the tensile Strength of Friction Stir Welded A356 Cast Aluminium Alloy by using Response Surface Methodology and Artificial Neural Network. It was observed that the error rate predicted by neural network was smaller than response surface methodology technique.

Facchini *et al* [4] developed a simulation model which was based on the adoption of the Artificial Neural Networks (ANNs) characterized by back-propagation learning algorithm with different types of architecture to predict of the Vickers Microhardness and Ultimate Tensile Strength of AA5754 H111 Friction Stir Welding Butt Joints.

Palanivel *et al* [5] predicted the tensile strength of dissimilar friction stir-welded AA6351–AA5083 using artificial neural network technique. In their study a feed-forward back propagation ANN architecture with a single hidden layer comprising 20 neurons was designed to simulate the ultimate tensile strength (UTS) of the joints. The models developed were capable of predicting values with less than 5 % error.

Tansel et al [6] used genetically optimized neural network systems (GONNS) to estimate the optimal operating condition of the friction stir welding (FSW) process. He introduced the genetically optimized neural network system (GONNS) by using Artificial Neural Network (ANN) and Genetic Algorithm (GA) together. He represented Friction Stir Welding (FSW) process in five artificial neural networks (ANN). Artificial Neural Network (ANN) is first trained by the genetically optimized neural network systems (GONNS) with experimental data. . It was observed that the inputs of the five ANNs were the same (tool rotation and welding feed rate). The estimation errors of the ANNs were better than average 0.5%. GA estimated the optimal FSW conditions to minimize or maximize one of the stir welding characteristics, while the others were kept at the desired ranges.

H. Okuyucu et al. [7] developed an artificial neural network (ANN) model for the analysis and simulation of the correlation between the friction stir welding (FSW) parameters of aluminium (Al) plates and mechanical properties. The input parameters of the model consist of weld speed and tool rotation speed (TRS). The outputs of the ANN model include property parameters namely: tensile strength, yield strength, elongation, hardness of weld metal and hardness of heat effected zone (HAZ). Good performance of the ANN model was achieved. The model can be used to calculate mechanical properties of welded Al plates as functions of weld & tool rotation speeds. The combined influence of weld speed and TRS on the mechanical properties of welded Al plates was simulated. A comparison was made between measured and calculated data. The calculated results were in good agreement with measured data. The aim of the paper was to show the possibility of the use of neural networks for the calculation of the mechanical properties of welded Al plates using FSW method. Results showed that, the networks can be used as an alternative in these systems.

L Fratini and G Buffa [8] studied the continuous dynamic re-crystallisation phenomena occurring in the FSW of Al alloys. A good agreement with the experimental results was obtained using the ANN model. In regard to ANNs, it noted that ANNs perform better than the other techniques, especially RSM when highly non-linear behaviour is the case. Also, this technique can build an efficient model using a small number of experiments; however the technique accuracy would be better when a larger number of experiments are used to develop a model.

In our present work we have used Quasi Newton algorithm to train the neural network architecture. The main objective is to determine the accuracy of the values of tensile strength and Vickers hardness which are obtained as outputs from the neural network.

2. Experimental Procedures

The alloy plates used in this case studies were AA6061 and AA7075 of dimensions 150 mm X 100 mm X 5 mm. The compositions of the both materials are shown in Table 1 and Table 2. H13 tool steel was used to Friction Stir Weld the dissimilar alloys in the butt configuration.

Chemical composition (wt.%)								
Mg	Si	Fe	Cu	Cr	Mn	Zn	Ti	Al
0.9	0.62	0.33	0.28	0.17	0.06	0.02	0.02	Rest

Table 1: Chemical Composition of AA 6061 alloys

Material	Weight Percent (%wt)							
	Zn	Cu	Mg	Si	Fe	Mn	Cr	Ti	Al
AA 7075-T651	6.2	1.7	2.7	0.1	0.2	0.1	0.2	0.02	Bal.

Table 2: Chemical Composition of AA 7075 alloys

The tensile test of the specimen shown in the Figure 1 was carried out on Universal Testing Machine (UTM). The experimental results are tabulated in the Table 3.

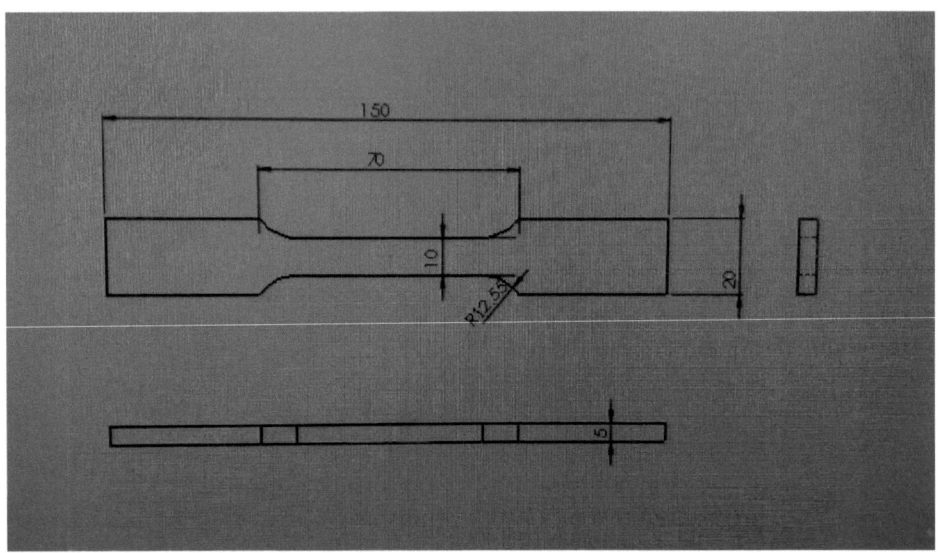

Figure 1: Tensile Test Specimen

Rotational Speed (rpm)	Traverse Speed (mm/min)	Ultimate Tensile Strength (MPa)	Hardness (HV)
800	6	212	79
1000	8	246	85
1200	10	248	102
1400	12	233	108
800	8	242	88
1000	6	248	82
1200	12	258	110
1400	10	255	98
800	10	270	96
1000	12	283	102
1200	6	265	83
1400	8	270	99
800	12	250	82
1000	10	275	80
1200	8	263	105
1400	6	245	83

Table 3: Experimental Values of Ultimate tensile strength (MPa) and Hardness (HV) at different Rotational Speed and Traverse speed

Artificial Neural Network architecture was constructed in Neural designer software. We imported the excel data sheet file upto the 15th experimental values consisting of 16 rows and 4 columns from the system to the software. The following flow chart depicts the procedure in the Figure 2:

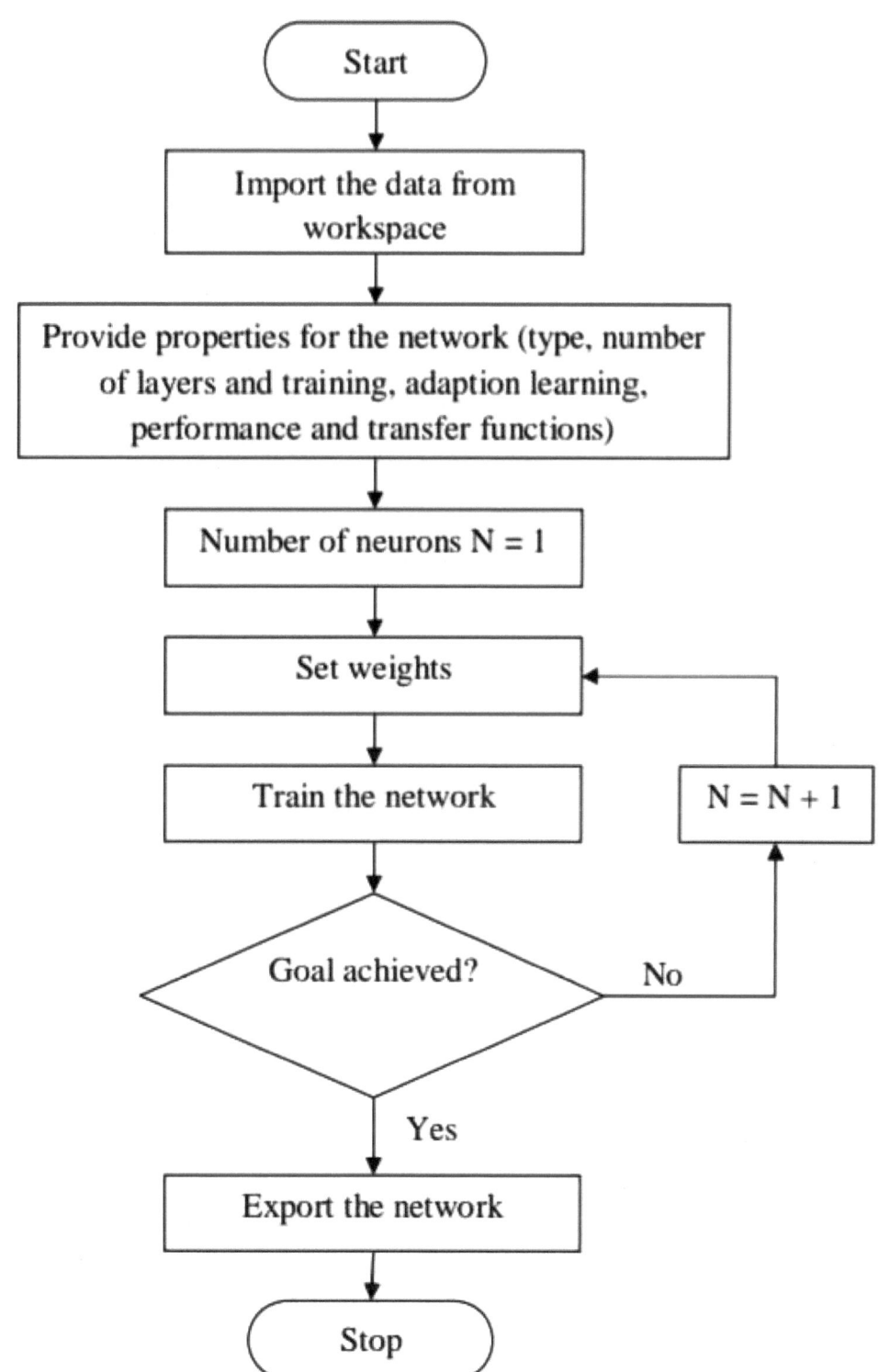

Figure 2: Flow chart of Artificial Neural Network architecture development

3. Results and discussions

3.1 Data Set

The data set contains the information for creating the predictive model. It comprises a data matrix in which columns represent variables and rows represent instances. Variables in a data set can be of three types: The inputs will be the independent variables; the targets will be the dependent variables; the unused variables will neither be used as inputs nor as targets. Additionally, instances can be: Training instances, which are used to construct the model; selection instances, which are used for selecting the optimal order; testing instances, which are used to validate the functioning of the model; unused instances, which are not used at all. The next table shows a preview of the data matrix contained in the file stir research technologies dissimilar al alloys.xlsx which was imported from the system to the software. Here, the number of variables is 4, and the number of instances is 15.

Rotational Speed (rpm)	Traverse Speed (mm/min)	Ultimate Tensile Strength (MPa)	Hardness (HV)
800	6	212	79
1000	8	246	85
1200	10	248	102
1400	12	233	108
800	8	242	88
1000	6	248	82
1200	12	258	110
1400	10	255	98
800	10	270	96
1000	12	283	102
1200	6	265	83
1400	8	270	99
800	12	250	82
1000	10	275	80
1200	8	263	105

Table 4: Data imported from the system to train the neural network

The following table depicts the names, units, descriptions and uses of all the variables in the data set. The numbers of inputs, targets and unused variables here are 2, 2, and 0, respectively.

	Name	Use
1.	Rotational Speed (rpm)	Input
2.	Traverse Speed (mm/min)	Input
3.	Ultimate Tensile Strength (MPa)	Output
4.	Hardness (HV)	Output

Table 5: Assignment of Inputs-Outputs in Neural Network Architecture

The next chart illustrates the variables use. It depicts the numbers of inputs (2), targets (2) and unused variables (0).

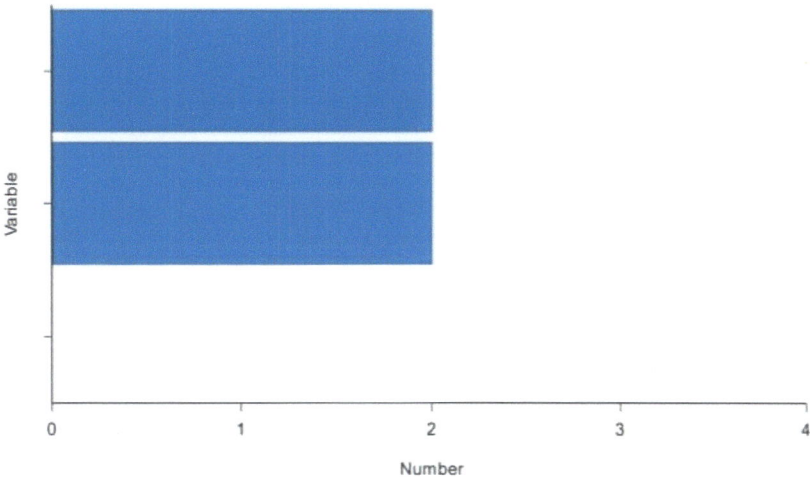

Figure 3: Variables bars chart

The following pie chart details the uses of all the instances in the data set. The total number of instances is 15. The number of training instances is 9 (60%), the number of selection instances is 3 (20%), the number of testing instances is 3 (20%), and the number of unused instances is 0 (0%).

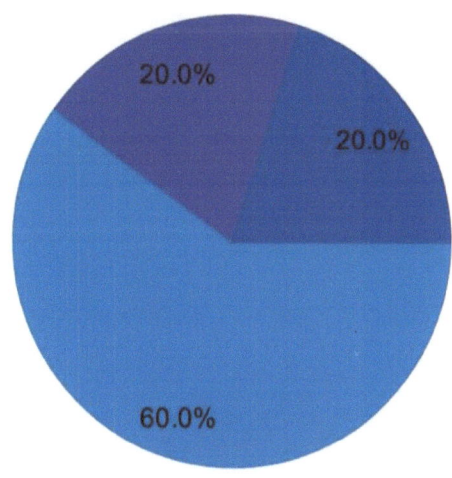

Figure 4: Instances Pie Chart

Basic statistics are a very valuable information when designing a model, since they might alert to the presence of spurious data. It is a must to check for the correctness of the most important statistical measures of every single variable. The table below shows the minimums, maximums, means and standard deviations of all the variables in the data set.

	Minimum	Maximum	Mean	Deviation
Rotational Speed (rpm)	800	1400	1080	224.245
Traverse Speed (mm/min)	6	12	9.2	2.24245
Ultimate Tensile Strength (MPa)	212	283	254.533	17.9478
Hardness (HV)	79	110	93.2667	10.9705

Table 6: Minimums, maximums, means and standard deviations of all the variables in the data set

Histograms show how the data is distributed over its entire range. In approximation problems, a uniform distribution for all the variables is, in general, desirable. If the data is very irregularly distributed, then the model will probably be of bad quality. The following chart shows the histogram for the variable Rotational Speed (rpm). The abscissa represents the centers of the containers, and the ordinate their corresponding frequencies. The minimum frequency is 0%, which corresponds to the bins with centers 890, 950, 1070, 1130, 1250 and 1310. The maximum frequency is 26.6667%, which corresponds to the bins with centers 830, 1010 and 1190.

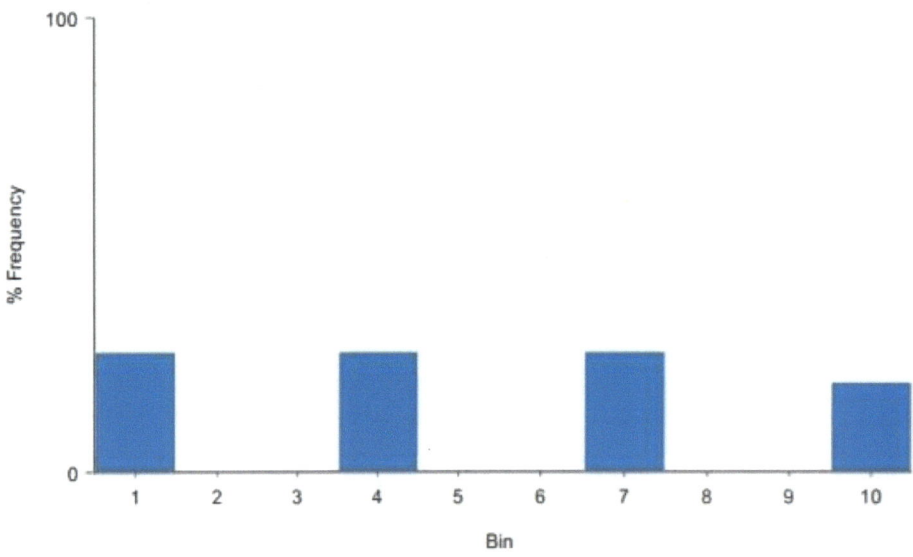

Figure 5: Rotational Speed (rpm) distribution

The following chart shows the histogram for the variable Welding Speed (mm/min). The abscissa represents the centers of the containers, and the ordinate their corresponding frequencies. The minimum frequency is 0%, which corresponds to the bins with centers 6.9,

7.5, 8.7, 9.3, 10.5 and 11.1. The maximum frequency is 26.6667%, which corresponds to the bins with centers 8.1, 9.9 and 11.7.

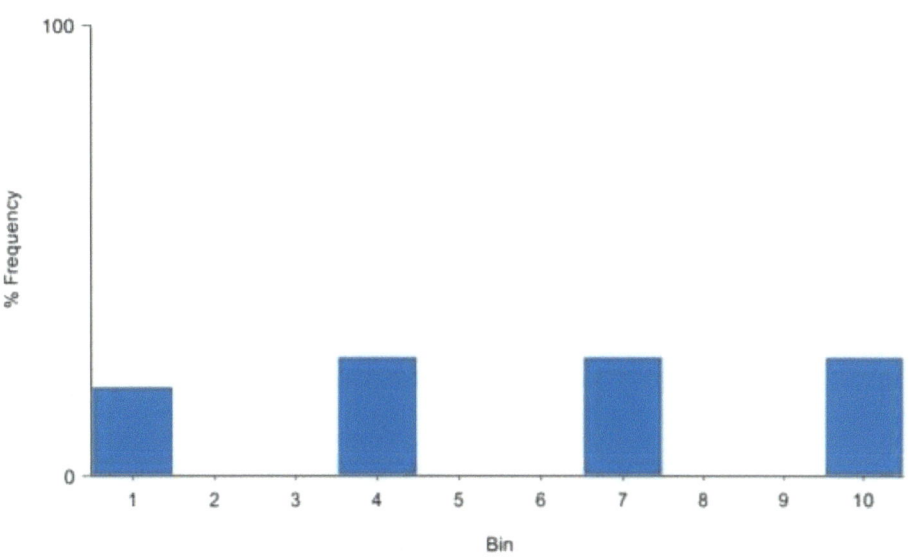

Figure 6: Welding Speed (mm/min) distribution

The following chart shows the histogram for the variable Ultimate Tensile Strength (Mpa). The abscissa represents the centers of the containers, and the ordinate their corresponding frequencies. The minimum frequency is 0%, which corresponds to the bins with centers 222.6 and 236.8. The maximum frequency is 20%, which corresponds to the bins with centers 251 and 272.4.

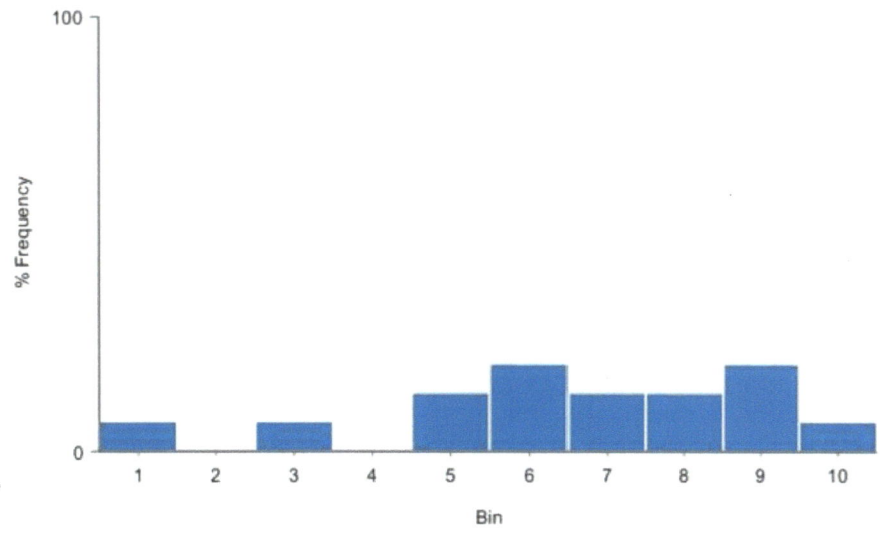

Figure 7: Ultimate Tensile Strength (Mpa) distribution

The following chart shows the histogram for the variable Hardness (HV). The abscissa represents the centers of the containers, and the ordinate their corresponding frequencies. The minimum frequency is 0%, which corresponds to the bins with centers 89.85 and 92.95. The maximum frequency is 26.6667%, which corresponds to the bin with center 80.55.

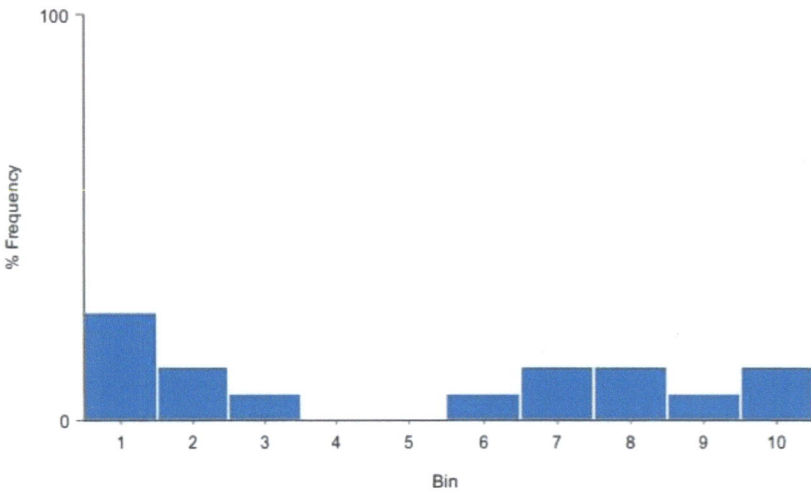

Figure 8: Hardness (HV) distribution

Box plots display information about the minimum, maximum, first quartile, second quartile or median and third quartile of every variable in the data set. They consist of two parts: a box and two whiskers. The length of the box represents the interquartile range (IQR), which is the distance between the third quartile and the first quartile. The middle half of the data falls inside the interquartile range. The whisker below the box shows the minimum of the variable while the whisker above the box shows the maximum of the variable. Within the box, it will also be drawn a line which represents the median of the variable. Box plots also provide information about the shape of the data. If most of the data are concentrated between the median and the maximum, the distribution is skewed right, if most of the data are concentrated between the median and the minimum, it is said that the distribution is skewed left and if there is the same number of values at the both sides of the median, the distribution is said to be symmetric.

The following chart shows the box plot for the variable Rotational Speed (rpm). The minimum of the variable is 800, the first quartile is 800, the second quartile or median is 1000, the third quartile is 1200 and the maximum is 1400.

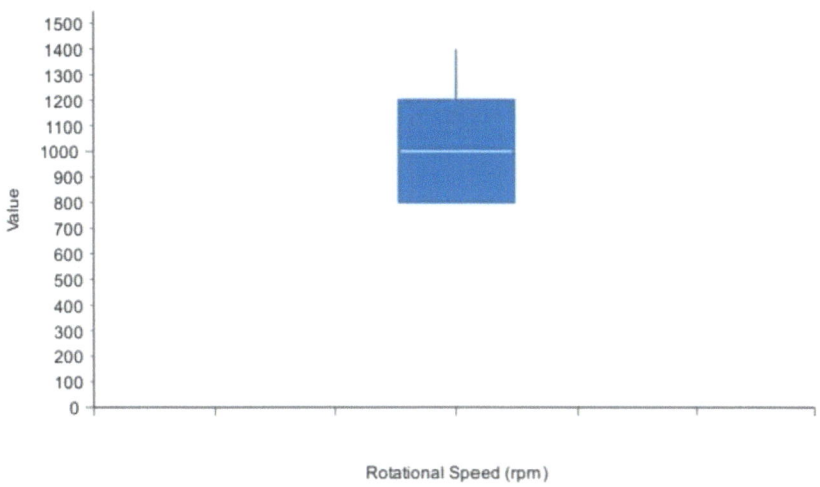

Figure 9: Rotational Speed (rpm) box plot

The following chart shows the box plot for the variable Welding Speed (mm/min). The minimum of the variable is 6, the first quartile is 8, the second quartile or median is 10, the third quartile is 12 and the maximum is 12.

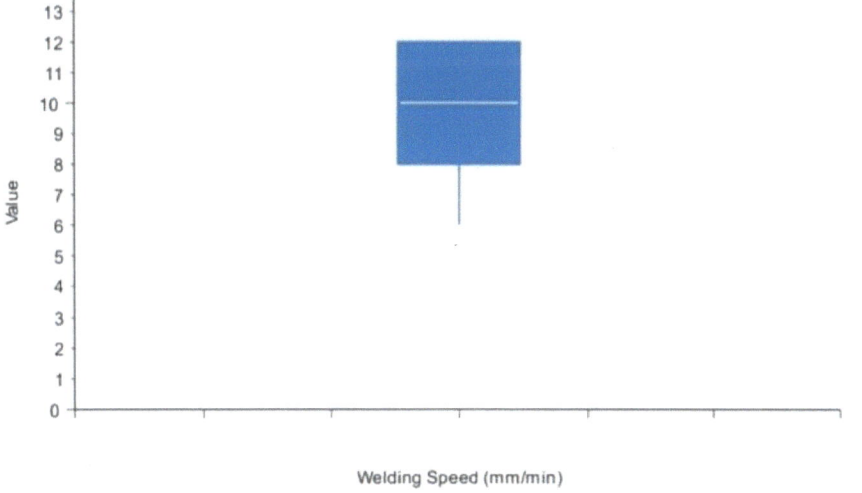

Figure 10: Welding Speed (mm/min) box plot

The following chart shows the box plot for the variable Ultimate Tensile Strength (Mpa). The minimum of the variable is 212, the first quartile is 246, the second quartile or median is 255, the third quartile is 270 and the maximum is 283.

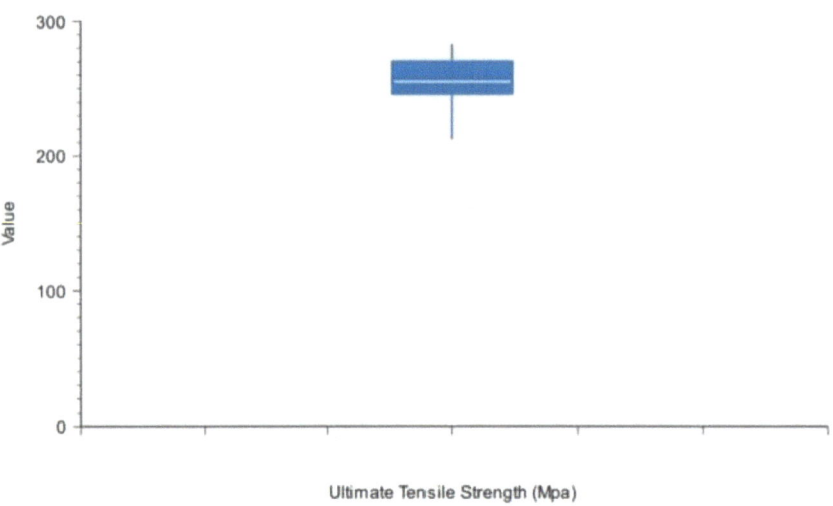

Figure 11: Ultimate Tensile Strength (Mpa) box plot

The following chart shows the box plot for the variable Hardness (HV). The minimum of the variable is 79, the first quartile is 82, the second quartile or median is 96, the third quartile is 102 and the maximum is 110.

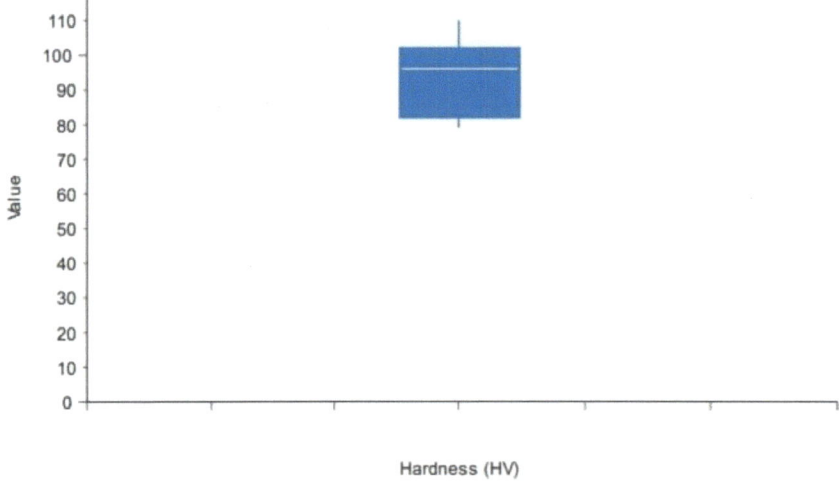

Figure 12: Hardness (HV) box plot

Target balancing task balances the distribution of targets in a data set for function regression. It unuses a given percentage of the instances whose values belong to the most populated bins. After this process, the distribution of the data will be more uniform and, in consequence, the resulting model will probably be of better quality.

The percentage of unused instances has been 10%, which corresponds to 1 instances. The following chart shows the histogram for the target variable Ultimate Tensile Strength (Mpa). The abscissa represents the centers of the containers, and the ordinate their corresponding frequencies. The minimum frequency is 0, which corresponds to the bins with centers 223 and 237. The maximum frequency is 3, which corresponds to the bin with center 272.

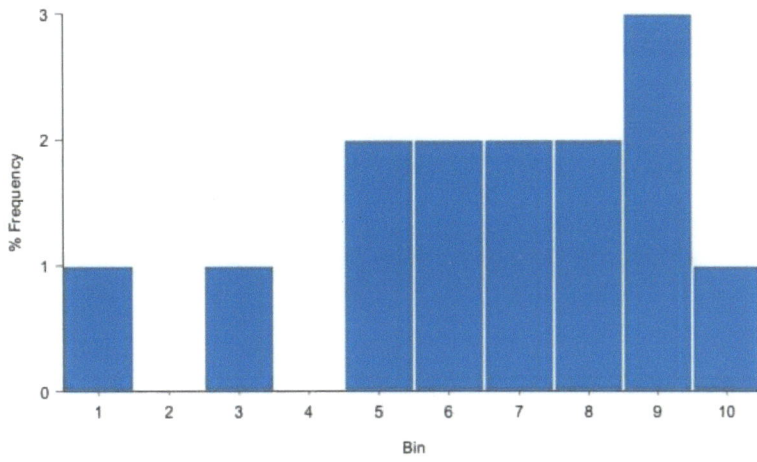

Figure 13: Ultimate Tensile Strength (Mpa) histogram

The percentage of unused instances has been 10%, which corresponds to 1 instances. The following chart shows the histogram for the target variable Hardness (HV). The abscissa represents the centers of the containers, and the ordinate their corresponding frequencies. The minimum frequency is 0, which corresponds to the bins with centers 89.8 and 92.9. The maximum frequency is 3, which corresponds to the bin with center 80.5.

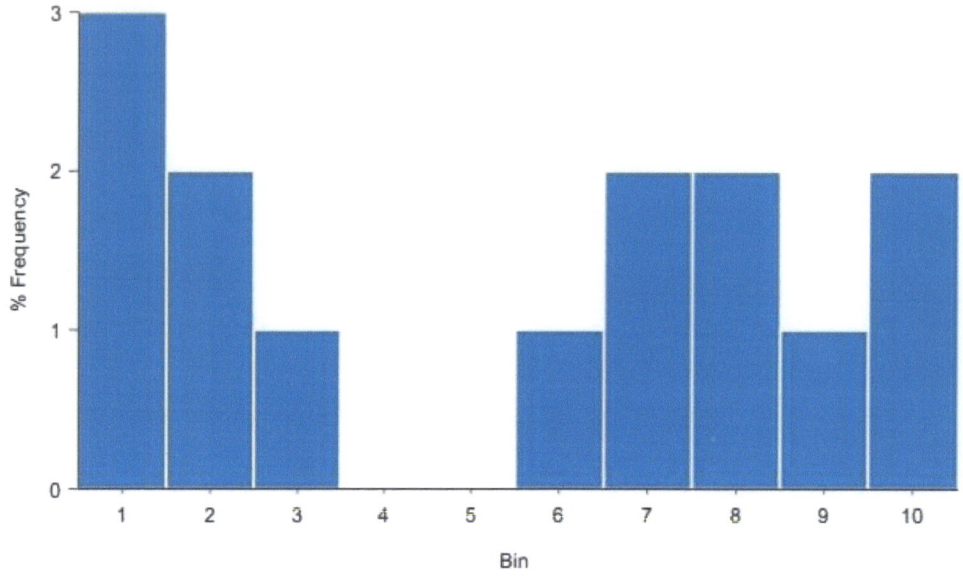

Figure 14: Hardness (HV) histogram

Scatter plot task plots graphs of all targets versus all input variables. That charts might help to see the dependencies of the targets with the inputs. The following chart shows the scatter plot for the input Rotational Speed (rpm) and the target Ultimate Tensile Strength (Mpa).

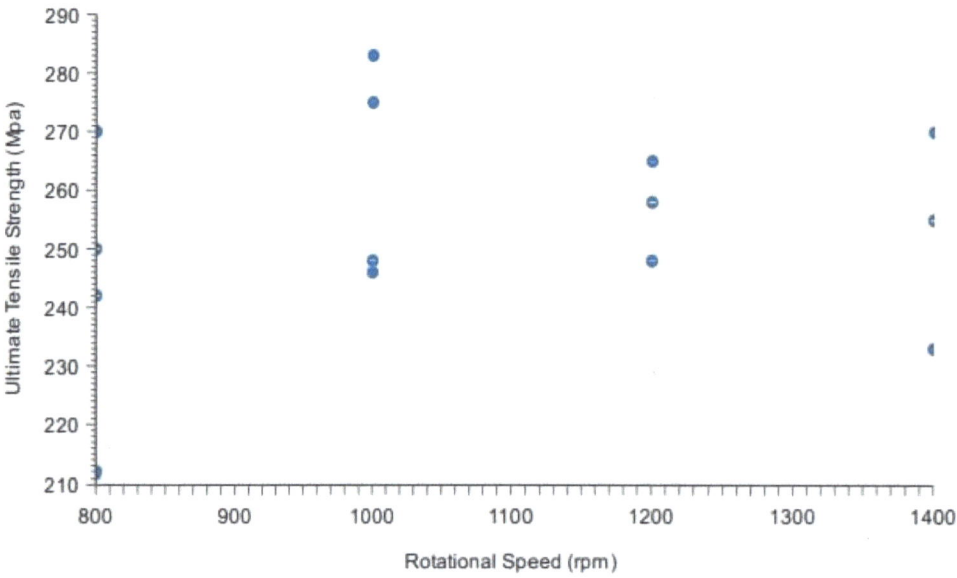

Figure 15: Ultimate Tensile Strength (Mpa) scatter chart vs Rotational Speed (rpm)

The following chart shows the scatter plot for the input Rotational Speed (rpm) and the target Hardness (HV).

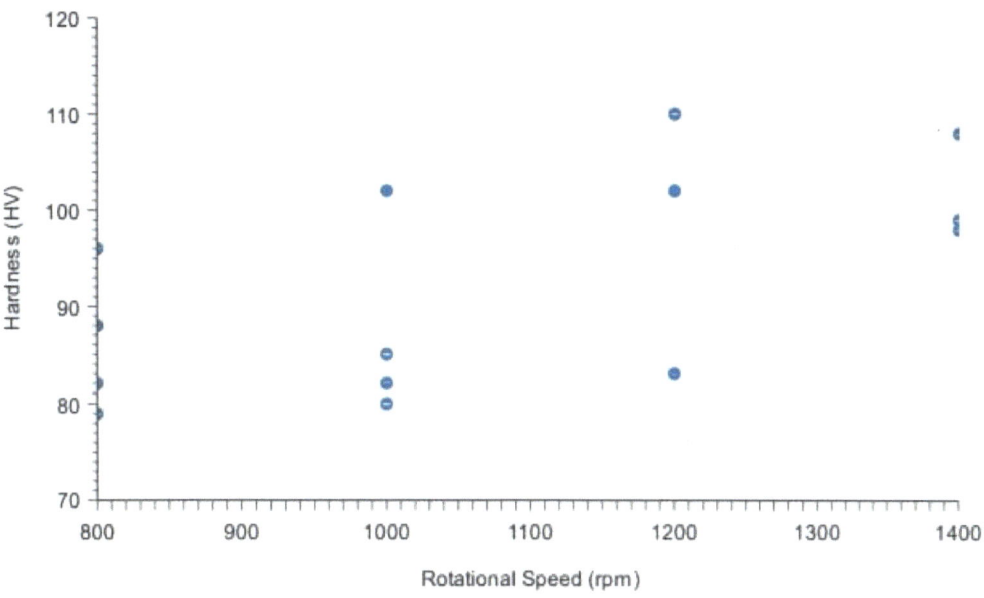

Figure 16: Hardness (HV) scatter chart vs Rotational Speed (rpm)

The following chart shows the scatter plot for the input Welding Speed (mm/min) and the target Ultimate Tensile Strength (Mpa).

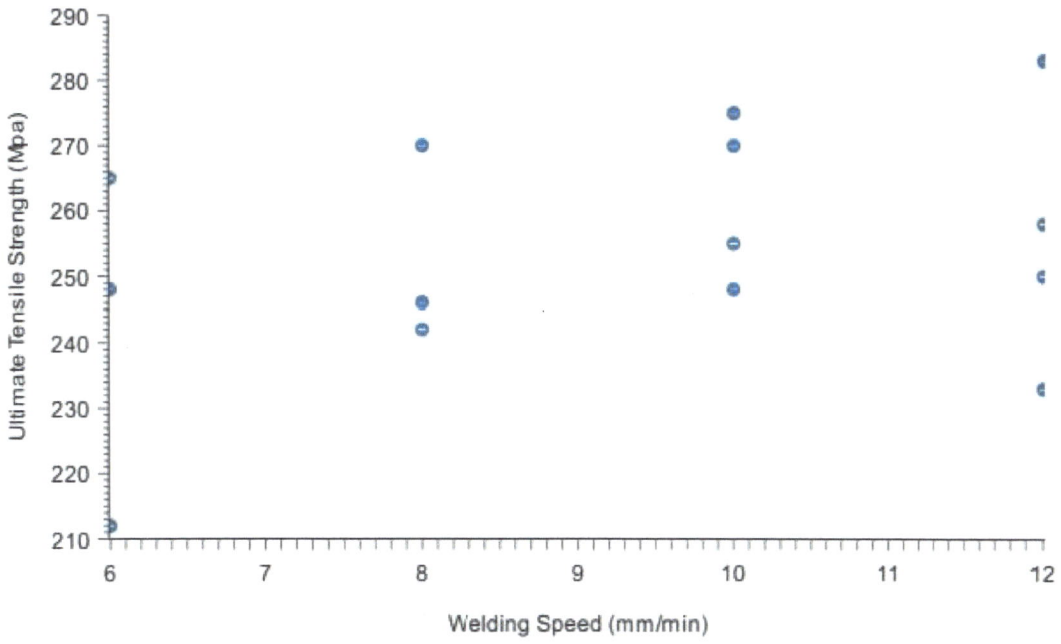

Figure 17: Ultimate Tensile Strength (Mpa) scatter chart vs Welding Speed (mm/min)

The following chart shows the scatter plot for the input Welding Speed (mm/min) and the target Hardness (HV).

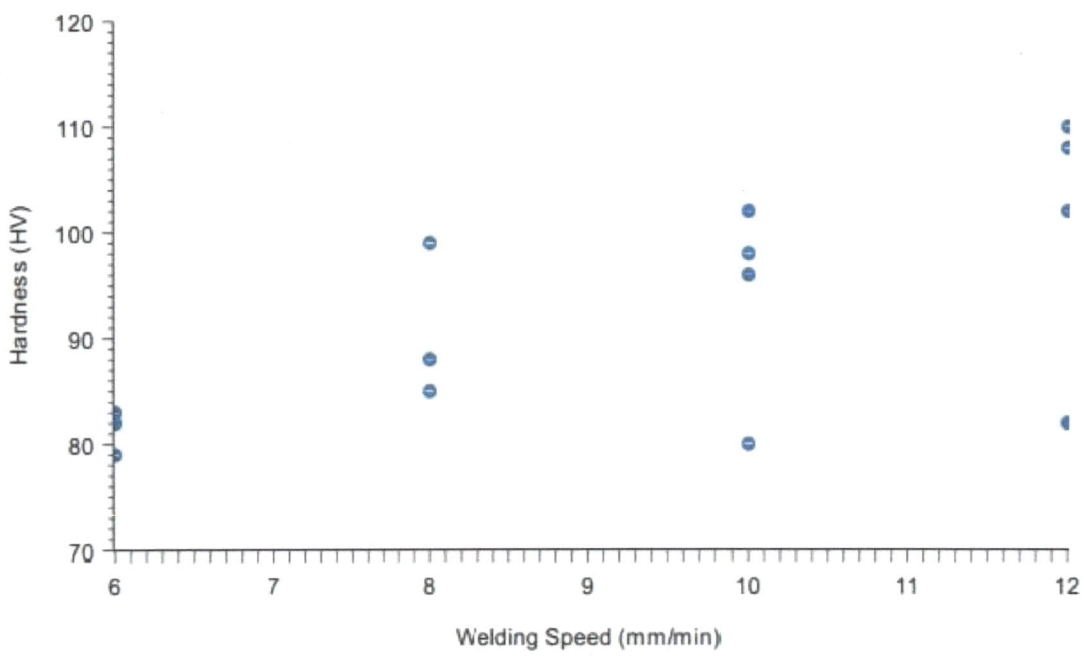

Figure 18: Hardness (HV) scatter chart vs Welding Speed (mm/min)

Correlation matrix task calculates the absolute values of the linear correlations among all inputs. The correlation is a numerical value between 0 and 1 that expresses the strength of the relationship between two variables. When it is close to 1 it indicates a strong relationship, and a value close to 0 indicates that there is no relationship. The following table shows the absolute value of the correlations between all input variables. The minimal correlation is 0.290323 between the variables Rotational Speed (rpm) and Welding Speed (mm/min). The maximal correlation is 0.290323 between the variables Rotational Speed (rpm) and Welding Speed (mm/min).

	Rotational Speed (rpm)	Welding speed (mm/min)
Rotational Speed (rpm)	1	0.29
Welding speed (mm/min)		1

Table 7: The absolute value of the correlations between all input variables

It might be interesting to look for linear dependencies between single input and single target variables. Linear correlations task calculates the absolute values of the correlation coefficient between all inputs and all targets. Correlations close to 1 mean that a single target is linearly correlated with a single input. Correlations close to 0 mean that there is not a linear relationship between an input and a target variable. Note that, in general, the targets depend on many inputs simultaneously. The following table shows the absolute value of the linear correlations between all input and target variables. The maximum correlation (0.612058) is yield between the input variable Rotational Speed (rpm) and the target variable Hardness (HV).

	Ultimate Tensile Strength (MPa)
Welding Speed (mm/min)	0.285
Rotational Speed (rpm)	0.173

Table 8: Ultimate Tensile Strength (Mpa) linear correlations

The next chart illustrates the dependency of the target Ultimate Tensile Strength (Mpa) with all the input variables.

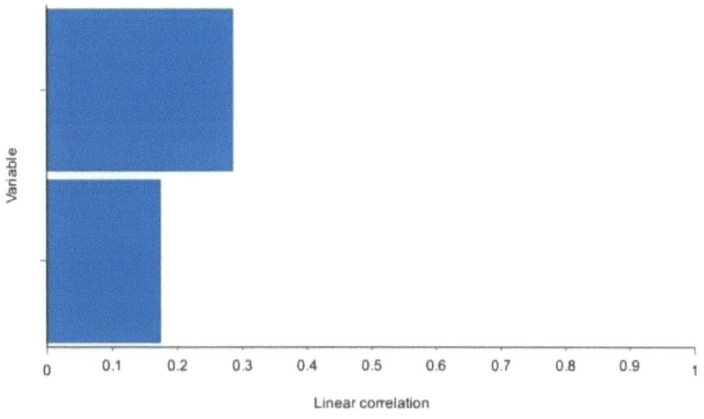

Figure 19: Ultimate Tensile Strength (Mpa) bars chart

The following table shows the absolute value of the linear correlations between all input and target variables. The maximum correlation (0.612058) is yield between the input variable Rotational Speed (rpm) and the target variable Hardness (HV).

	Hardness (HV)
Rotational Speed (rpm)	0.612
Welding Speed (mm/min)	0.561

Table 9: Hardness (HV) linear correlations

The next chart illustrates the dependency of the target Hardness (HV) with all the input variables.

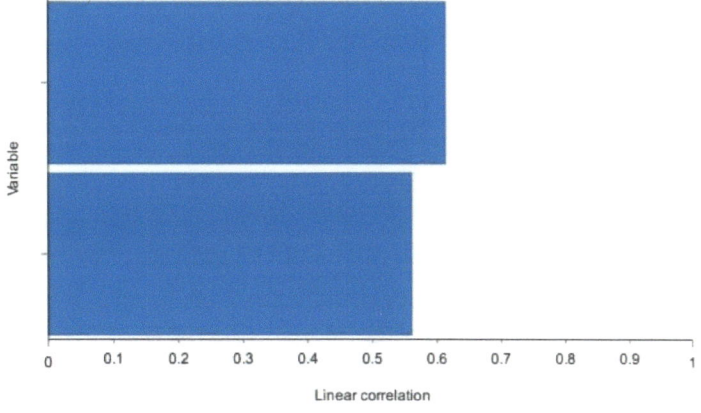

Figure 20: Hardness (HV) bars chart

When designing a predictive model, the general practice is to first divide the data into three subsets. The first subset is the training set, which is used for constructing different candidate models. The second subset is the selection set, which is used to select the model exhibiting the best properties. The third subset is the testing set, which it is used for validating the final model. The following table shows the uses of all the instances in the data set. Note that the instances are arranged in rows of 10. The total number of instances is 15. The numbers of training, selection, testing and unused instances are 10, 2, 2 and 1, respectively.

	1	2	3	4	5	6	7	8	9	10
0	Test	Train	Train	Sel.	Sel.	Train	Train	Train	Train	Train
10	Train	Train	Unused	Test	Train					

Table 10: Instances table

The following pie chart details the uses of all the instances in the data set. There are 10 instances for training (66.7%), 2 instances for selection (13.3%), 2 instances for testing (13.3%) and 1 unused instances (6.67%).

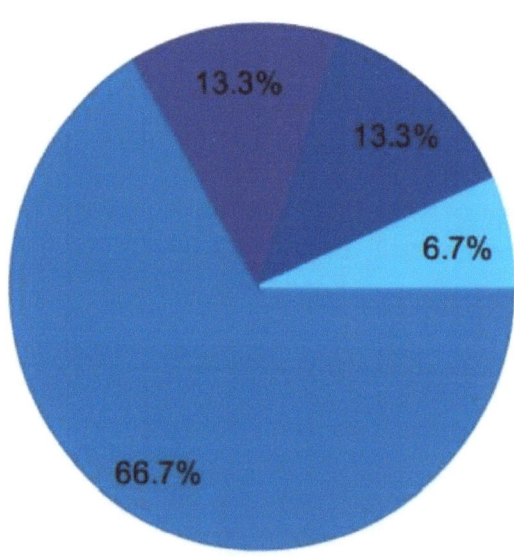

Figure 21: Instances pie chart

Principal components analysis allows to identify underlying patterns in a data set so it can be expressed in terms of other data set of lower dimension without much loss of information. The resulting data set should be able to explain most of the variance of the original data set by making a variable reduction. The final variables will be named principal components. Since this process is not reversible, it will be only applied to the input variables. The next table shows in the first column the relative explained variance for every of the principal components and in the second column the cumulative explained variance. The number of principal components of the resulting data set depends on the minimum value of the cumulative variance that it is desired the final data set had.

	Relative Variance	Cumulative Variance
1	64.5161	64.5161
2	35.4839	100

Table 11: Principal components results

The next chart shows the cumulative explained variance for the principal components. The x-axis represents each of the principal components and the y-axis depicts the the cumulative explained variance. As it can be seen, the total explained variance for all the principal components is 100% but if the number of chosen principal components drecreases also makes it the total explained variance.

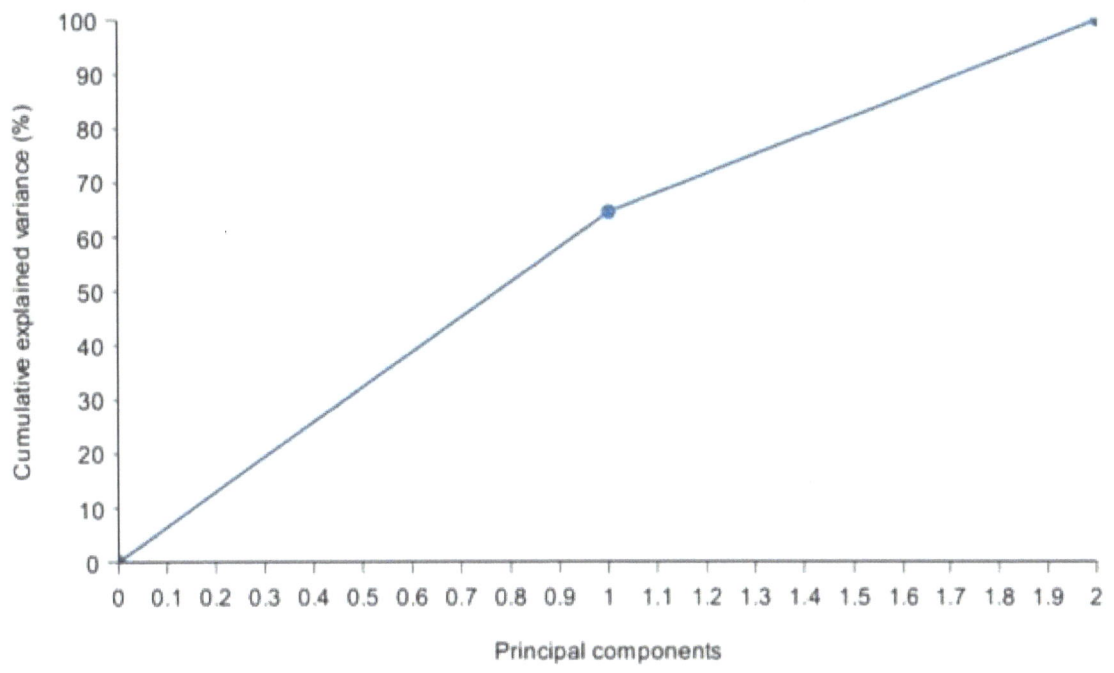

Figure 22: Explained variance chart

3.2 Neural Network

The neural networks represent the predictive model. In Neural Designer neural networks allow deep architectures, which are a class of universal approximator. The size of the scaling layer is 2, the number of inputs. The scaling method for this layer is the Minimum-Maximum. The following table shows the values which are used for scaling the inputs, which include the minimum, maximum, mean and standard deviation.

	Minimum	Maximum	Mean	Deviation
Rotational Speed (rpm)	800	1.4e+003	1.08e+003	224
Welding Speed (mm/min)	6	12	9.2	2.24

Table 12: Scaling layer

The number of layers in the neural network is 2. The following table depicts the size of each layer and its corresponding activation function. The architecture of this neural network can be written as 2:3:2.

	Input number	Neurons number	Activation Function
1	2	3	Hyperbolic Tangent
2	3	2	Linear

Table 13: Size of each layer and its corresponding activation function

The following table shows the statistics of the parameters of the neural network. The total number of parameters is 17.

	Minimum	Maximum	Mean	Standard Deviation
Statistics	-1.98	0.871	-0.205	0.859

Table 14: Neural network parameters

The size of the unscaling layer is 2, the number of outputs. The unscaling method for this layer is the minimum and maximum. The following table shows the values which are used for scaling the inputs, which include the minimum, maximum, mean and standard deviation.

	Minimum	Maximum	Mean	Deviation
Ultimate Tensile Strength (MPa)	212	283	255	18.6
Hardness (HV)	79	110	94.1	10.9

Table 15: The values which are used for scaling the inputs

A graphical representation of the network architecture is depicted next. It contains a scaling layer, a neural network and an unscaling layer. The yellow circles represent scaling neurons, the green circles the principal components, the blue circles perceptron neurons and the red circles unscaling neurons. The number of inputs is 2, the number of principal components is 2, and the number of outputs is 2. The complexity, represented by the numbers of hidden neurons, is 3.

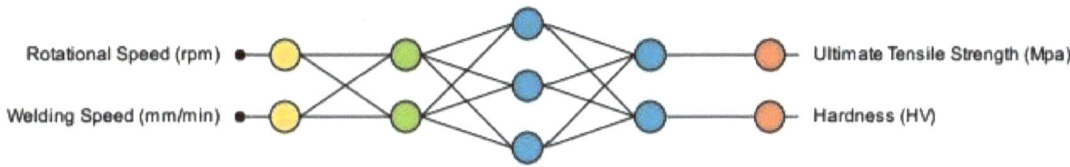

Figure 23: Neural Network Architecture

The loss index plays an important role in the use of a neural network. It defines the task the neural network is required to do, and provides a measure of the quality of the representation that it is required to learn. The choice of a suitable loss index depends on the particular application. The normalized squared error is used here as the error method. It divides the squared error between the outputs from the neural network and the targets in the data set by a normalization coefficient. If the normalized squared error has a value of unity then the neural network is predicting the data 'in the mean', while a value of zero means perfect prediction of the data. The neural parameters norm is used as the regularization method. It is applied to control the complexity of the neural network by reducing the value of the parameters. The weight of this regularization term in the loss expression is 0.001.

The procedure used to carry out the learning process is called training (or learning) strategy. The training strategy is applied to the neural network in order to obtain the best possible loss. The quasi-Newton method is used here as training algorithm. It is based on Newton's method, but does not require calculation of second derivatives. Instead, the quasi-Newton method computes an approximation of the inverse Hessian at each iteration of the algorithm, by only using gradient information. The following plot shows the losses in each iteration. The initial value of the training loss is 1.13974, and the final value after 65 iterations is 0.0794535. The initial value of the selection loss is 4.48423, and the final value after 65 iterations is 20797.6.

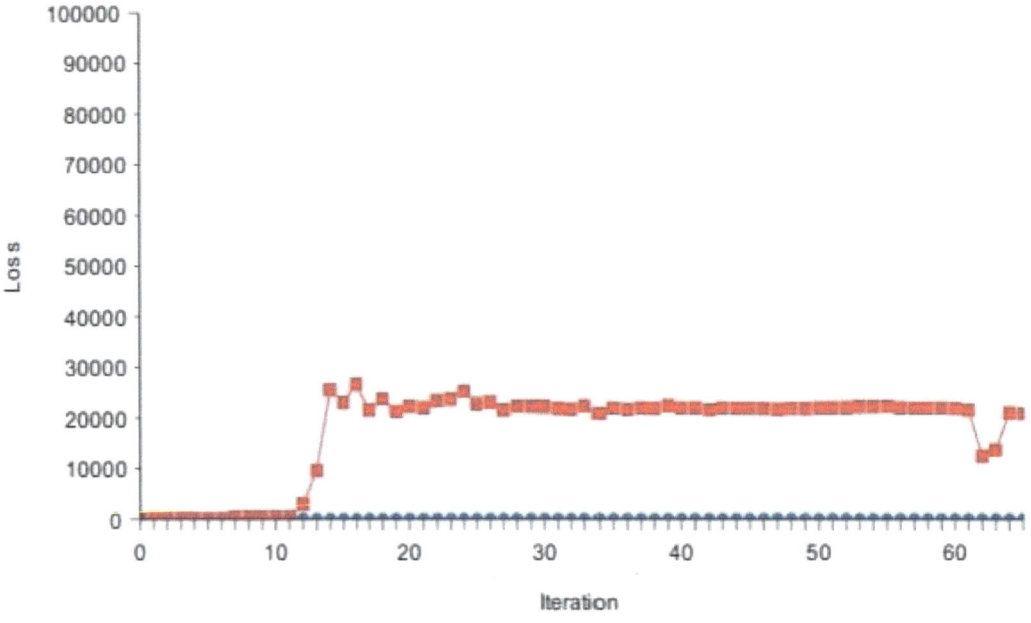

Figure 24: Quasi-Newton method losses history

The next table shows the training results by the quasi-Newton method. They include some final states from the neural network, the loss functional and the training algorithm.

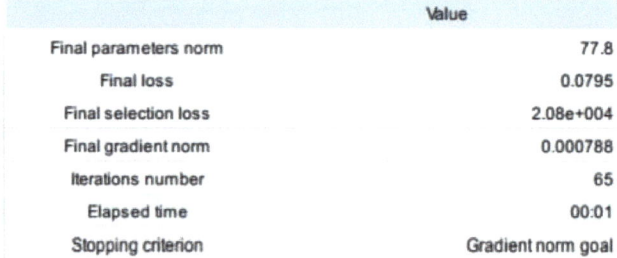

	Value
Final parameters norm	77.8
Final loss	0.0795
Final selection loss	2.08e+004
Final gradient norm	0.000788
Iterations number	65
Elapsed time	00:01
Stopping criterion	Gradient norm goal

Table 16: Quasi Newton method Results

3.3 Model Selection

Model selection is applied to find a neural network with a topology that optimizes the loss on new data. There are two different types of algorithms for model selection: Order selection algorithms and input selection algorithms. Order selection algorithms are used to find the optimal number of hidden neurons in the network. Inputs selection algorithms are responsible for finding the optimal subset of input variables.

Inputs importance tasks calculates the selection loss when removing one input at a time. This shows which input have more influence in the outputs. The next table shows the importance of each input. If the importance takes a value greater than 1 for an input, it means that the selection error without that input is greater than with it. In the case that the importance is lower than 1, the selection error is lower without using that input. Finally, if the importance is

1, there is no difference between using the current input and not using it. The most important variable is Rotational Speed (rpm), that gets a contribution of 85.75% to the outputs.

	Contribution
Rotational Speed (rpm)	85.75
Welding Speed (mm/min)	1.45

Table 17: Inputs importance results

The next chart illustrates the contribution of each input.

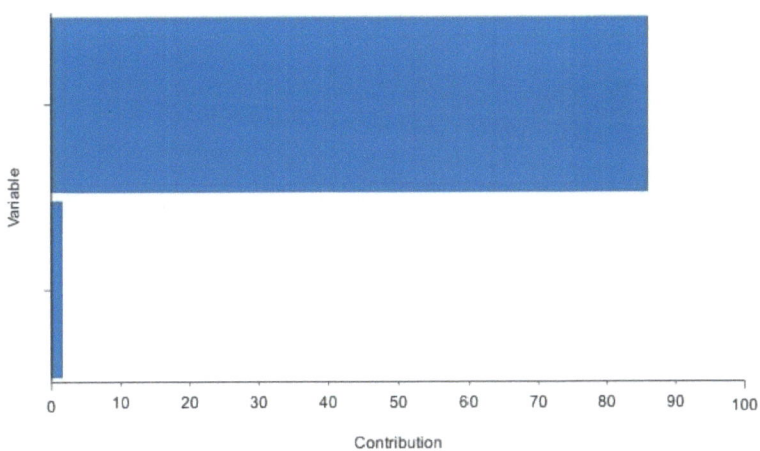

Figure 25: Contribution of each input

The best selection is achieved by using a model whose complexity is the most appropriate to produce an adequate fit of the data. The order selection algorithm is responsible of finding the optimal number of neurons in the network. Incremental order is used here as order selection algorithm in the model selection. The next chart shows the loss history for the different subsets during the incremental order selection process. The blue line represents the training loss and the red line symbolizes the selection loss.

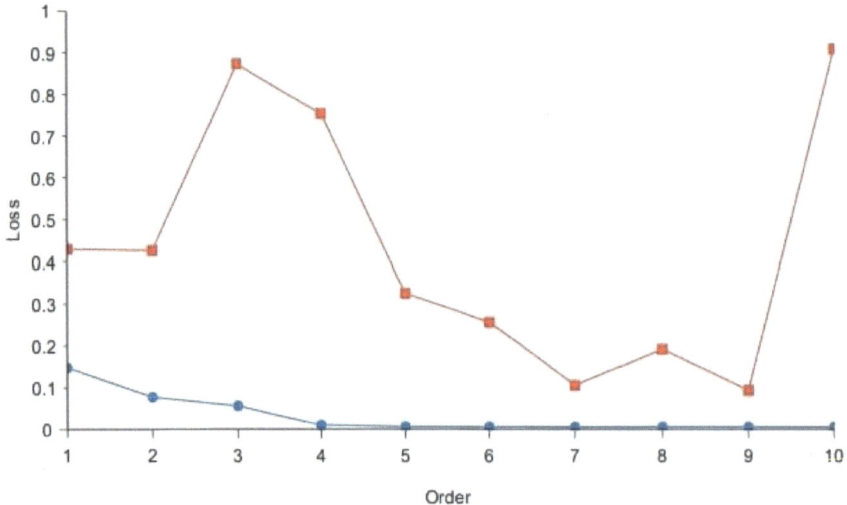

Figure 26: Incremental order losses plot

The next table shows the order selection results by the incremental order algorithm. They include some final states from the neural network, the loss functional and the order selection algorithm.

	Value
Optimal order	9
Optimum training loss	0.00590464
Optimum selection loss	0.0933822
Iterations number	10
Elapsed time	00:15

Table 18: Incremental order results

A graphical representation of the resulted deep architecture is depicted next. It contains a scaling layer, a neural network and an unscaling layer. The yellow circles represent scaling neurons, the green circles the principal components, the blue circles perceptron neurons and the red circles unscaling neurons. The number of inputs is 2, the number of principal components is 2, and the number of outputs is 2. The complexity, represented by the numbers of hidden neurons, is 9.

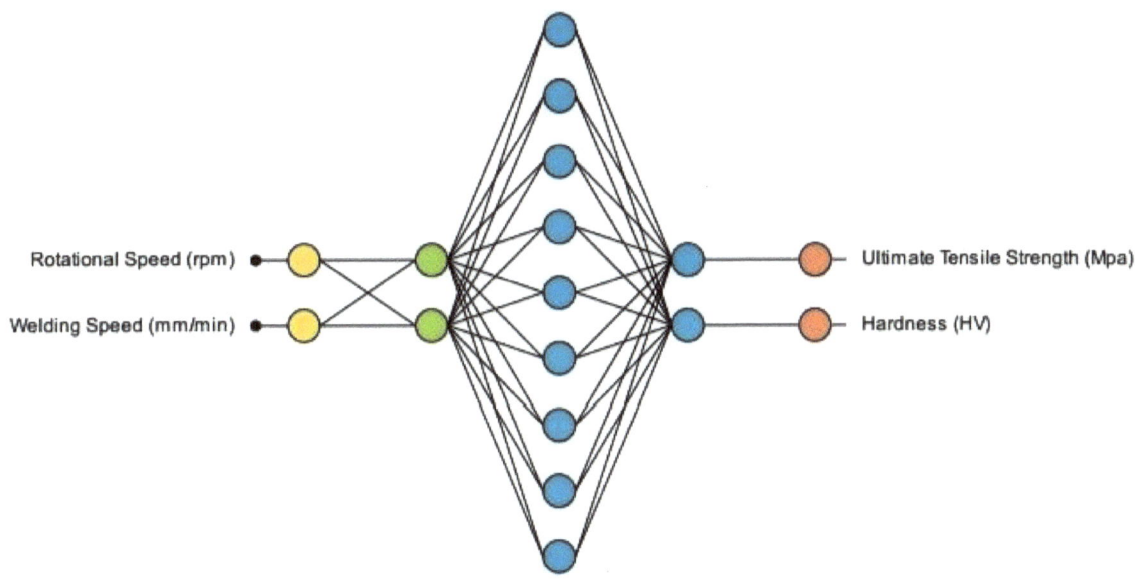

Figure 27: Final Architecture

A neural network produces a set of outputs for each set of inputs applied. The outputs depend, in turn, on the values of the parameters. The next table shows the input values and their corresponding output values. The input variables are Rotational Speed (rpm) and Welding Speed (mm/min); and the output variables are Ultimate Tensile Strength (Mpa) and Hardness (HV).

	Value
Rotational Speed (rpm)	1400
Welding Speed (mm/min)	6
Ultimate Tensile Strength (Mpa)	249.722951
Hardness (HV)	69.3800818

Table 19: Inputs-outputs table

It is very useful to see the how the outputs vary as a function of a single input, when all the others are fixed. This can be seen as the cut of the neural network model along some input direction and through some reference point. The next table shows the reference point for the plots.

	Value
Rotational Speed (rpm)	1400
Welding Speed (mm/min)	6

Table 20: Reference point table

The next plot shows the output Ultimate Tensile Strength (Mpa) as a function of the input Rotational Speed (rpm). The x and y axes are defined by the range of the variables Rotational Speed (rpm) and Ultimate Tensile Strength (Mpa), respectively.

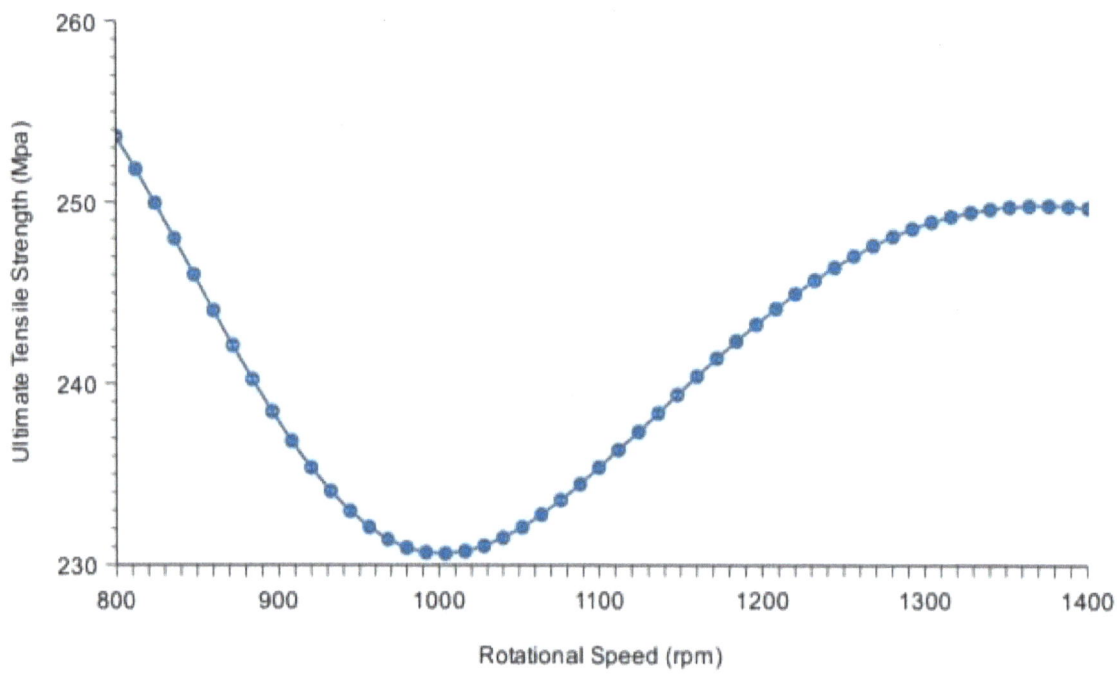

Figure 28: Ultimate Tensile Strength (Mpa) against Rotational Speed (rpm) directional line chart

The next plot shows the output Hardness (HV) as a function of the input Rotational Speed (rpm). The x and y axes are defined by the range of the variables Rotational Speed (rpm) and Hardness (HV), respectively. Note that some directional outputs fall outside the range of Hardness (HV), and therefore they are not plotted.

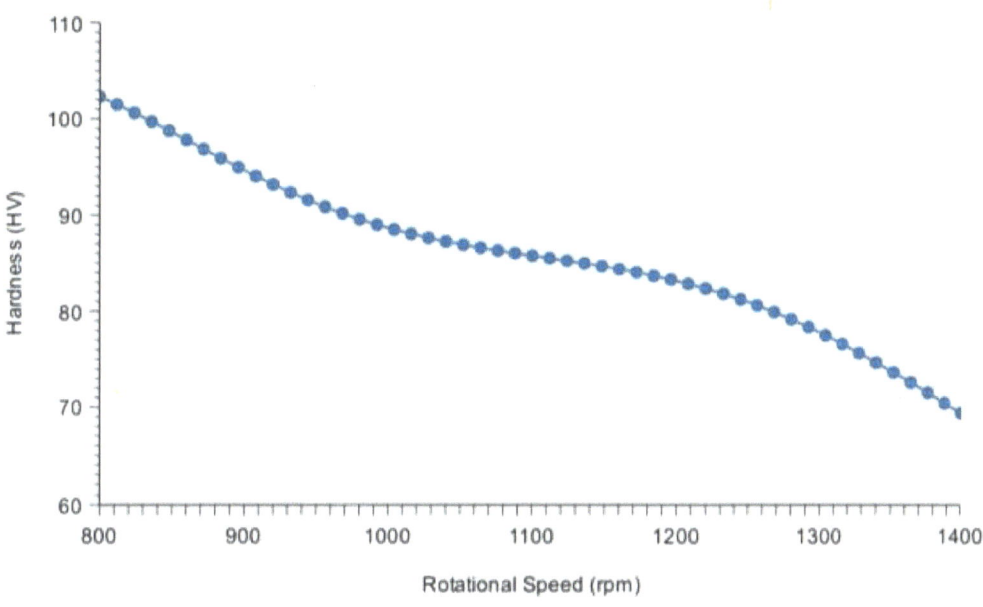

Figure 29: Hardness (HV) against Rotational Speed (rpm) directional line chart

The next plot shows the output Ultimate Tensile Strength (Mpa) as a function of the input Welding Speed (mm/min). The x and y axes are defined by the range of the variables Welding Speed (mm/min) and Ultimate Tensile Strength (Mpa), respectively.

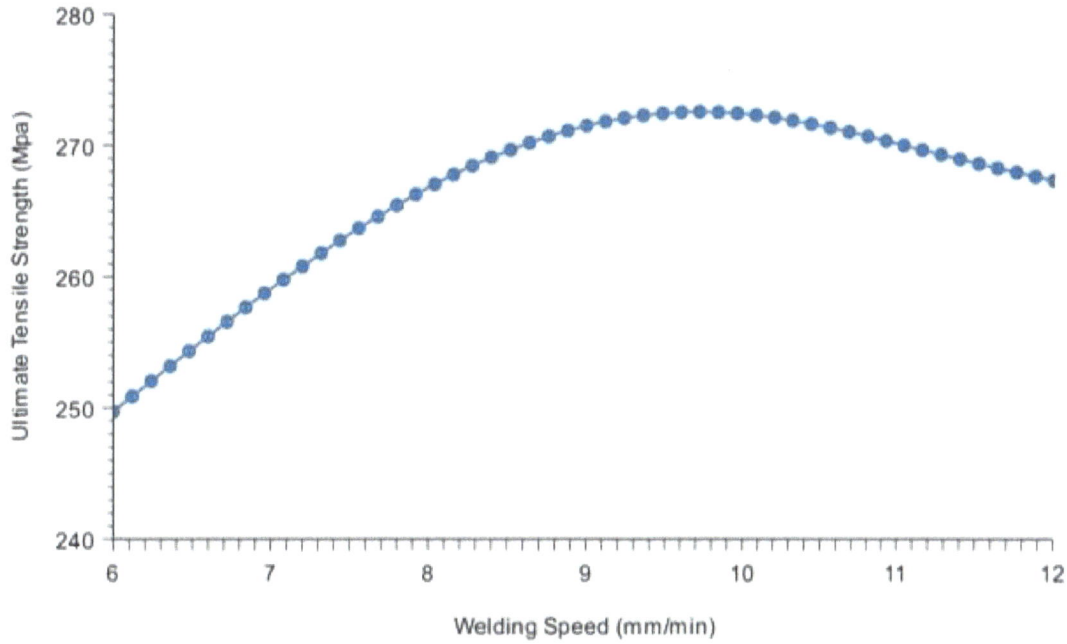

Figure 30: Ultimate Tensile Strength (Mpa) against Welding Speed (mm/min) directional line chart

The next plot shows the output Hardness (HV) as a function of the input Welding Speed (mm/min). The x and y axes are defined by the range of the variables Welding Speed (mm/min) and Hardness (HV), respectively. Note that some directional outputs fall outside the range of Hardness (HV), and therefore they are not plotted.

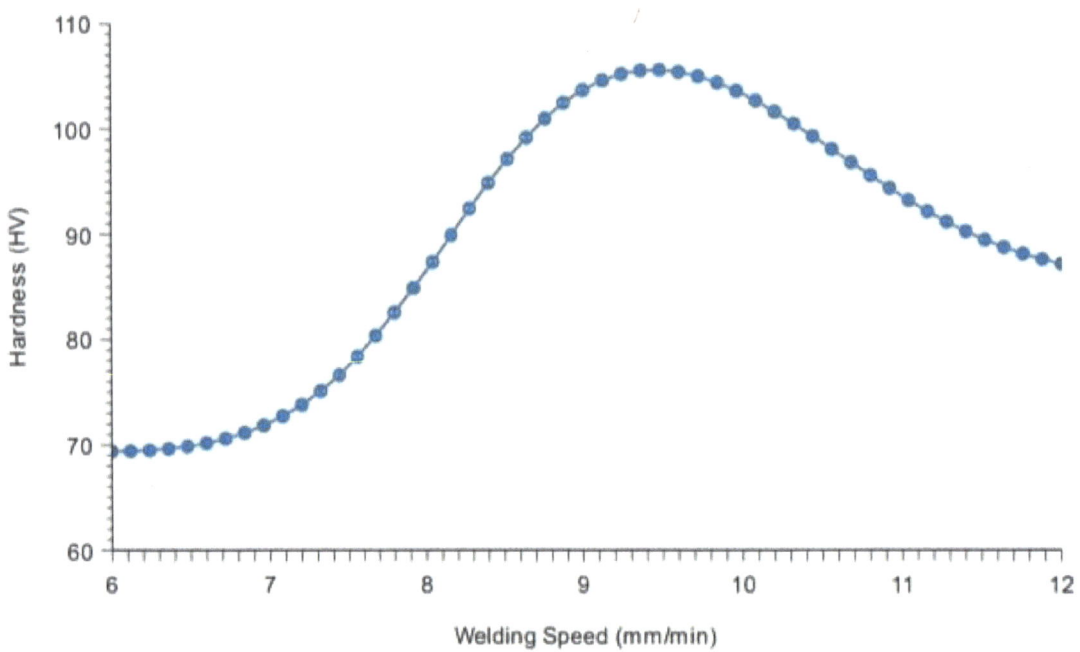

Figure 31: Hardness (HV) against Welding Speed (mm/min) directional line chart

4. Conclusion

The actual tensile strength and Vickers hardness calculated at the rotational speed of 1400 rpm and 6 mm/ min traverse speed is 245 MPa and 83 HV while predicted tensile strength and Vickers hardness from Neural Network architecture trained on Quasi Newton algorithm are 249.72 MPa and 69.38 HV. So the accuracy for predicting the tensile strength and Vickers hardness using the neural network architecture trained on Quasi Newton algorithm is 98.07% and 83.59%. It can be concluded that Neural Network architecture can be used to reduce cost and time of the experiment.

References

1. Zhang, Hong-Chao & Huang, Samuel. (1995). Applications of neural networks in manufacturing: A state-of-the-art survey. International Journal of Production Research - INT J PROD RES. 33. 705-728. 10.1080/00207549508930175.

2. 2. Shojaeefard, M.H., Khalkhali, A., Akbari, M. and Asadi, P., 2015. Investigation of friction stir welding tool parameters using FEM and neural network. Proceedings of the Institution of Mechanical Engineers, Part L: Journal of Materials: Design and Applications, 229(3), pp.209-217.

3. Jayaraman, M., Sivasubramanian, R., Balasubramanian, V. and Lakshminarayanan, A.K., 2008. Prediction of tensile strength of friction stir welded A356 cast aluminium alloy using response surface methodology and artificial neural network. Journal for Manufacturing Science and Production, 9(1-2), pp.45-60.

4. De Filippis, L., Serio, L., Facchini, F., Mummolo, G. and Ludovico, A., 2016. Prediction of the vickers microhardness and ultimate tensile strength of AA5754 H111 friction stir welding butt joints using artificial neural network. Materials, 9(11), p.915.

5. Palanivel, R., Laubscher, R.F., Dinaharan, I. and Murugan, N., 2016. Tensile strength prediction of dissimilar friction stir-welded AA6351–AA5083 using artificial neural network technique. Journal of the Brazilian Society of Mechanical Sciences and Engineering, 38(6), pp.1647-1657.

6. Tansel, Ibrahim N., et al. "Optimizations of friction stir welding of aluminum alloy by using genetically optimized neural network." The International Journal of Advanced Manufacturing Technology 48.1-4 (2010): 95-101.

7. Hasan Okuyucu a, Adem Kurt a, Erol Arcaklioglu -Artificial neural network application to the friction stir welding of aluminum plates- Materials and Design 28 (2007) 78–84

8. L Fratini and G Buffa- Continuous dynamic recrystallization phenomena modelling in friction stir welding Proceedings of the Institution of Mechanical Engineers; May 2007; 221, B5; ProQuest Science Journals pg. 857

www.ingramcontent.com/pod-product-compliance
Lightning Source LLC
Chambersburg PA
CBHW041313180526
45172CB00004B/1085